THE ARCTIC JOURNALS
OF JOHN RAE

Selected and introduced by
KEN McGOOGAN

Introduction, selection, and notes copyright © 2012 Ken McGoogan

All rights reserved. No part of this publication may be reproduced, stored in a retrieval system, or transmitted in any form or by any means—electronic, mechanical, recording, or otherwise—without the prior written consent of the publisher or a licence from The Canadian Copyright Licensing Agency (ACCESS Copyright). For a copyright licence, visit accesscopyright.ca.

TouchWood Editions
touchwoodeditions.com

LIBRARY AND ARCHIVES CANADA CATALOGUING IN PUBLICATION
Rae, John, 1813–1893
The Arctic journals of John Rae / introduction by
Ken McGoogan.

Issued also in electronic formats.
ISBN 978-1-927129-74-6

1. Rae, John, 1813–1893—Travel—Arctic regions.
2. Rae, John, 1813–1893—Travel—Northwest, Canadian.
3. Arctic regions—Description and travel. 4. Northwest, Canadian—Description and travel. 5. Inuit—Canada—Social life and customs—19th century. 6. Scientific expeditions. 7. Ethnological expeditions. I. Title.

FC3961.R34 2012   917.1904'1   C2012-902646-8

Proofreader: Elizabeth McLachlin
Cover image: William Armstrong, "Dr. John Rae (1913–1893), Arctic Explorer (Full Length Portrait, 1862)," watercolour on off-white paper, Collection of Glenbow Museum, Calgary, Canada, 55.17.1

We gratefully acknowledge the financial support for our publishing activities from the Government of Canada through the Canada Book Fund, Canada Council for the Arts, and the province of British Columbia through the British Columbia Arts Council and the Book Publishing Tax Credit.

The interior pages of this book have been printed on 100% post-consumer recycled paper, processed chlorine free, and printed with vegetable-based inks.

1 2 3 4 5 16 15 14 13 12

PRINTED IN CANADA

Thinking from what I saw that I should
like the wild sort of life . . .
>   —John Rae

CONTENTS

❈ ❈ ❈

Map of Northwest Passage in 1845 —⚞— vi
Introduction by Ken McGoogan —⚞— 1

Part One: The Lost Autobiography —⚞— 9
Part Two: The Making of an Arctic Explorer —⚞— 45
Part Three: John Rae Contends with Charles Dickens —⚞— 237

About John Rae —⚞— 311
About Ken McGoogan —⚞— 312

The Northwest Passage Region as known in 1845, when the Franklin Expedition sailed.

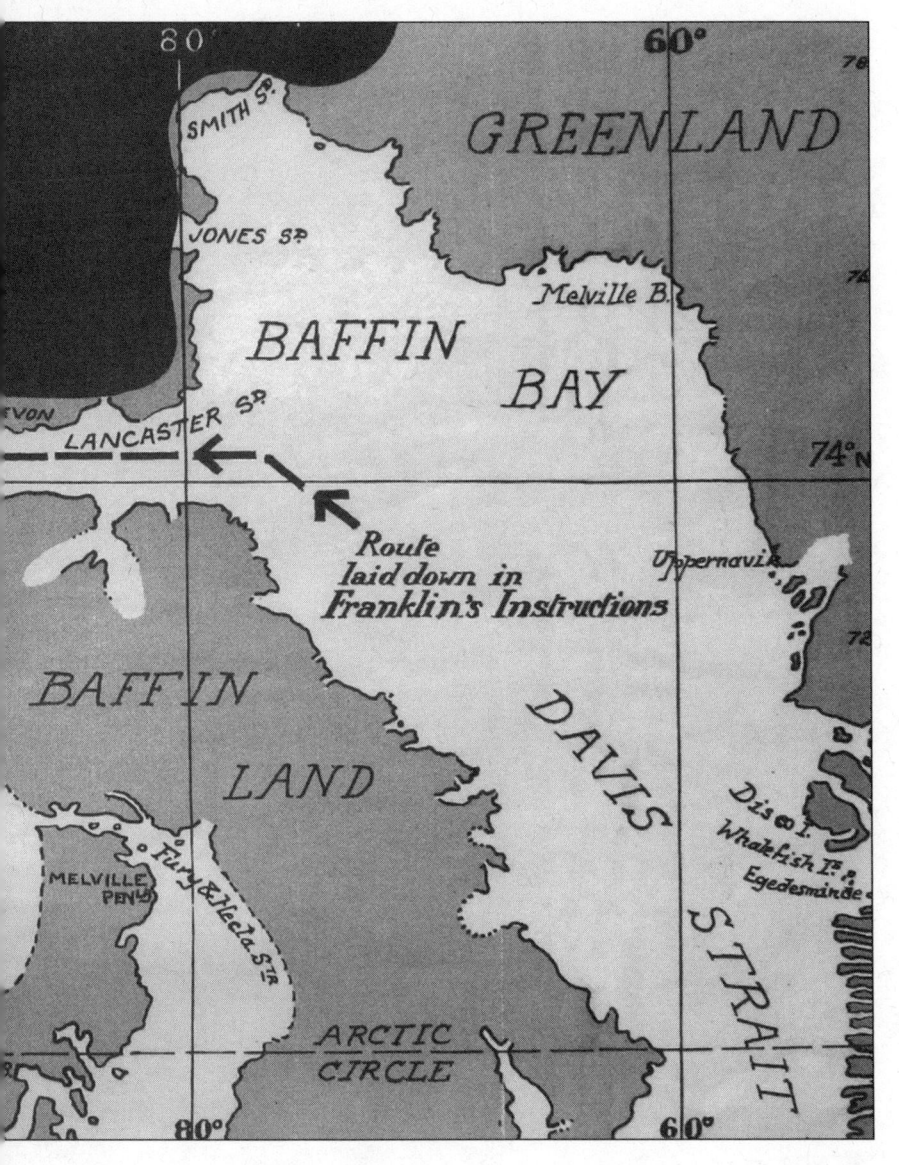

INTRODUCTION

The handwritten, nine-hundred-page autobiography of John Rae ends abruptly in mid-sentence, just as the explorer is about to glean the first inkling of the disaster that had engulfed the lost expedition of Sir John Franklin. Rae describes how, in April 1854, he was slogging west across Boothia Peninsula in the High Arctic, beating his way through heavy snow. "It was impossible to keep a straight course," he writes, "and we had to turn to the northward out of our course, so as to select the . . ."

So ends the manuscript, with Rae about to meet a small group of Inuit hunters and hear allegations that would excite Victorian England to frenzy. But what happened to the rest of the autobiography? Scholars have been wondering since 1968, when Alan Cooke, curator at the Scott Polar Research Institute (SPRI) in Cambridge, England, expressed both pride at acquiring the document and "intense disappointment" at the way it ended. "What became of the rest of the manuscript?" Cooke wondered. "Certainly once there was more. After recounting in full flow the events of so many years, Rae would not have found it possible to stop just short of the greatest discovery of his career."

The mid-sentence ending invited speculation. Some readers suggested that Rae might have suppressed the remainder of the manuscript because it revealed some dark secret he had decided to bury. The mystery remained unsolved until 1998, when I spent three months in Cambridge, scouring the files at SPRI for what would become *Fatal Passage: The Untold Story of John Rae, the Arctic Explorer Who Discovered the Fate of Franklin*.

My researches carried me into a lavishly illustrated, large-format work by D. Murray Smith entitled *Arctic Expeditions from British and Foreign Shores / From the Earliest Times to the Expedition of 1875–76*. That work has appeared in several editions under the same title. Luckily, I was reading the eight-hundred-and-twenty-four-page version published by Charles H. Calvert in 1877. In that work, while discussing Rae's final expedition, Smith wrote that the explorer had provided him with as-yet-unpublished material: "It is with great pleasure, therefore, that the present writer is enabled to present to the public the following notes, embracing fresh particulars in connection with his most interesting episode in Arctic Exploration. These valuable notes have, in the kindest manner, been supplied by Dr. Rae for the present work. We give Dr. Rae's communication, so courteously sent, in the form in which it has come to hand."

John Rae's "valuable notes," published here for the first time since 1877, included the long-lost final pages of the explorer's autobiography. And to this we shall return.

John Rae (1813–1893) solved the two great mysteries of nineteenth-century Arctic exploration. He discovered both the final link in the Northwest Passage and the fate of the 1845 expedition led by Sir John Franklin. These accomplishments capped a career during which, between 1846 and 1854, drawing on methods and techniques gleaned from the native peoples, he led four major Arctic expeditions.

In 1846–47, Rae became the first explorer to spend a winter above the Arctic Circle while living off the land. In 1848–49, while officially second-in-command to John Richardson (who left halfway through), Rae explored the Arctic coast from the Mackenzie River to the Coppermine. In 1850–51, he conducted what is arguably the most remarkable sledge-and-boat expedition in Arctic exploration history. And in 1853–54, during his final expedition, he solved the two mysteries mentioned above. Rae's revelations regarding the Franklin expedition—that its final survivors had resorted to cannibalism—shook Victorian England to its foundations. To sort that out, I had to write a whole second book: *Lady Franklin's Revenge*.

The present volume, *The Arctic Journals of John Rae*, is divided into three sections. "Part One: The Lost Autobiography" comprises two and a half chapters from D. Murray Smith's *Arctic Expeditions*. While theoretically accessible to Arctic aficionados, this 1870s perspective on Rae has never been available to a broad reading public. In "Part VII" of his book, Smith devotes one chapter to Rae's expedition of 1846–47. That chapter is not included here because the present volume contains Rae's entire journal from that expedition.

Later, in "Part XII: The Fate of Franklin Ascertained," Smith dedicates two and a half chapters to Rae. The first of these begins with the heading "Dr. Rae's autobiography." It provides a good overview of Rae's career, outlining his background as an Orcadian Scot, an Edinburgh-trained doctor, and a Hudson's Bay Company officer, and summarizing his second and third expeditions. Smith devotes the next one and a half chapters to "Rae's Last Arctic Expedition." After drawing a few pages from the explorer's report to the Hudson's Bay Company, written at York Factory and dated September 1, 1854, Smith quotes Rae at length.

But here we need more context. At the Scott Polar Research Institute, Rae's papers include a note from his wife, Katherine, who reported that her husband "had a great dislike to writing, and though continually urged by myself and my sister and friends to write his autobiography, we could seldom get him to work steadily at it. He would often return to what he had written from time to time and tear up parts & make additions to others which will account for the different dates of the notes, and for the fragmentary style." Anyone who spends time with Rae's autobiographical palimpsest, which includes scientific notes and explanations running to fifteen pages, will recognize this as understatement.

Long story short: while working on his autobiography, Rae must have got wind of Smith's *Arctic Expeditions*, or perhaps even read an early edition. Rather than recopy the final pages of his manuscript, he gathered them together with his published reports, added a few introductory sentences, and sent the whole package to Smith. More than a century later, while researching *Fatal Passage*, I was unable to determine what had happened to Smith's papers. If they have survived, Rae's handwritten pages will almost certainly be among them.

"Part Two" of the present volume, "The Making of an Arctic Explorer," comprises and reprints John Rae's only published book, *Narrative of an expedition to the shores of the Arctic Sea, in 1846 and 1847*. In November 1847, three years before the book appeared in print, Rae returned to Britain on leave from the HBC and published a brief report of his expedition in *The London Times*.

Sir John Richardson of the Royal Navy had recently agreed to lead an overland search for Sir John Franklin, who had disappeared into the Arctic with two ships in 1845. Now fifty-nine years of age and serving as a hospital administrator, Richardson—a leading naturalist, but no great traveller—had not set foot in the north for twenty years. He was seeking a second-in-command who would compensate for his deficiencies. When he read Rae's brief report in *The Times*, he jumped to his feet and cried out to his wife, "I have found my companion, if I can get him."

John Rae was a thirty-four-year-old doctor who had worked in the fur trade for fourteen years. Energetic, resourceful, an avid practitioner of Inuit travel methods, Rae had thrived for months in one of the most extreme environments on the planet. What so impressed Richardson in the newspaper report stands fully revealed in the published *Narrative*. The SPRI curator Cooke, a well-informed analyst, noted Rae's foresight, practicality, incisive originality, and "contempt for orthodox opinion and incompetence." In these pages, as well, we discover an inspirational leader of almost superhuman endurance.

On his 1846–47 expedition, Rae charted 655 miles of unknown coastline and demonstrated that Boothia Peninsula was not an island, as widely supposed. Eventually, on his four Arctic expeditions, Rae would survey 1,765 miles of uncharted territory while travelling 6,555 miles on snowshoes and sailing 6,700 miles in small boats. On this first expedition, as on every one he later led, Rae served as chief hunter and supplier of food. Here, too, he stood alone among his contemporaries.

Within *Narrative*, Rae's prowess as an outdoorsman gives rise to entertaining moments. My favorite is the one when, out hunting deer before daybreak, he spots a group of animals moving rapidly toward him. These turn out to be seventeen wolves. They race forward to within forty yards, when they break off into a semi-circle. Rae drops to one knee, takes aim at the leader, and fires. In the grey light he merely grazes the wolf, cutting off a line of hair and skin.

"They apparently did not expect to meet with such a reception," he tells us, "for after looking at me a second or two they trotted off, no doubt as much disappointed at not making a breakfast of me as I was at missing my aim. Had they come to close quarters [which they sometimes do when pressed for food], I had a large and strong knife which would have proved a very efficient weapon."

In 1848, when with Richardson he departed to search for Franklin, Rae left his handwritten *Narrative* with a friend, who had undertaken to see it through the publishing process. This friend was in no hurry, Rae wrote later, "for my manuscript was sent by request to the hydrographer [chief mapmaker] of the Admiralty and allowed to remain more than a year in his hands, although he had no right to have it, and the little book did not make its appearance until 1850."

By that time, Rae added, interest had dwindled. What's more, some editor had taken a sharp pencil to the manuscript, which "had been so remodelled that I did not know my own bantling." Anyone who compares the clear, readable *Narrative* with Rae's digression-filled autobiography will conclude that this "remodelling" was probably for the best. (The attentive reader will note a few repetitions in the present volume, as when Rae describes his discovery of certain Franklin relics. These repetitions have been allowed to stand for the sake of coherence and completeness.)

"Part Three" of this book, "John Rae Contends with Charles Dickens," focuses on the controversy that erupted in Victorian England after Rae returned with the first news of the fate of the Franklin expedition. Basically, Rae sent a brief, upbeat letter to *The Times*. To the Admiralty, he wrote a more detailed report. This latter report, which he did not intend for publication, ended up on the front page of *The Times*. Lady Franklin, the widow of Sir John, was outraged. To repudiate Rae, she enlisted the aid of Charles Dickens, the most influential writer of the age, and also the publisher of a weekly journal, *Household Words*.

Of the six documents collected in "Part Three" for the first time, two were written by Dickens. Without these, Rae's rebuttal would remain incomprehensible. Not surprisingly, Dickens produced an eloquent tour de force. But as I wrote in *Fatal Passage*, his two-part essay was also "a masterpiece of obfuscation, self-deception, and almost wilful blindness." More than that, some passages in the two-part screed reveal a profound racism, and must be seen to be believed. Consider, for example, the passage that begins: "The word of a savage is not to be taken for it [cannibalism]; firstly because he is a liar; secondly because he is a boaster; thirdly because he talks figuratively . . ."

By the end, Dickens stands revealed in such an unflattering light that, in the acclaimed docudrama *Passage*, which is based on *Fatal Passage*, one of the author's twenty-first-century descendants (Gerald Dickens) offers a public apology to the Inuit. As for the dispute over cannibalism, and the refusal by some to believe that it occurred, contemporary researchers have vindicated John Rae and his Inuit informants in every particular.

To his credit, Charles Dickens concluded the exchange by publishing Rae's original, 1854 report to the Hudson's Bay Company, and that is the final document included here. In it, Rae describes eventually travelling north along the west coast of Boothia Peninsula with just two men, an Inuk and an Ojibwe. He discovers that King William Land is "clearly an island"—one separated from Boothia Peninsula by a previously unknown strait. Shortly after coming to this realization, he writes, "having taken possession of our discoveries in the usual form, and built a cairn, we commenced our return."

John Rae had discovered the final link in the Northwest Passage, the only channel navigable by ships of that era. Yet he would remain the only great British explorer never to receive a knighthood. Indeed, he has yet to be accorded proper recognition. Not long ago, an Orkney-based member of the British House of Commons moved a motion to have Rae commemorated in Westminster Abbey, where a bust of Sir John Franklin can be found. That motion went down in flames.

But the battle continues. In May 2012, as this book was in preparation, *The Literary Review of Canada* published an article by Adrian Craciun, who questioned the wisdom of devoting a national historic site, location yet to be determined, to the long-lost Franklin expedition—"a failed British expedition," she wrote, "whose architects sought to demonstrate the superiority of British science over Inuit knowledge."

Invited to respond, I could only elaborate. Instead of obsessing over two Royal Navy ships that disappeared in 1845, I suggested that Canadians should create a national historic site on Boothia Peninsula at the spot—specified in the final pages of this book—where John Rae built a cairn overlooking the strait that would later bear his name. Forget about searching for two long-lost ships, and commemorating what Craciun rightly called "the tragic uselessness" of the Franklin disaster. We should focus instead on building a site to mark the discovery—by the Scot, the Ojibwe, and the Inuk—of the final link in the Northwest Passage.

Ken McGoogan
July 2012
Toronto, Ontario

PART ONE

THE LOST AUTOBIOGRAPHY

An excerpt from *Arctic Expeditions from British and Foreign Shores / From the Earliest Times to the Expedition of 1875–76* by D. Murray Smith, featuring material quoted from Rae's notes and journals.

CHAPTER ONE

Dr. Rae's Autobiography—A Hunting Adventure—Boat-Building Under Difficulties—Coast Of Wollaston Land Explored—Coast Of Victoria Land Explored.

A few weeks after the return of Captain Belcher and his disappointed and discomfited officers and crews, all Europe and America were profoundly affected by the intelligence that the fate of Sir John Franklin's party had at last been ascertained. Englishmen had not been satisfied with the achievements of Sir Edward Belcher, who had abandoned the search for the missing expedition immediately after it had been clearly demonstrated that the "Erebus" and "Terror," or at least relics of these good ships, *would be found* within a certain well-defined and strictly circumscribed area to the west of Boothia Peninsula. The whole of the known Polar world between Baffin's Bay and Behring Strait had been searched, *except* this limited area, and it seemed unaccountable that the commander of four distinct crews, all amply provisioned, and each capable of undertaking the thorough examination of thousands of miles of coast line in a single season, should, at the moment when the course of events at last pointed out the one district now left unsearched, have ordered the total and the final abandonment of the enterprise which had employed the best efforts of the Navy for ten years, and which now seemed so near a happy attainment. For the abandonment of the Franklin search was indeed final. The expedition had been lost for nine years, and it was the opinion of many of those best qualified to judge, that no colony of Europeans, dependent on their own exertions exclusively, could support life for so long a time in a region so rigorous in climate and so barren in productions. It was understood, therefore, that the Lords Commissioners of the Admiralty had agreed to consider the further prosecution of the search as hopeless. Probably the outbreak of the Crimean War in 1854 had something to do with their apparent unanimity on this point. However this may be, it was now evident to the friends and relatives of our missing countrymen, that in any future measures that might be undertaken with the view of clearing up the inscrutable and most painful mystery, the Government were not likely to take part. But just at this time, when many Englishmen were reluctantly making up their minds to rest content in ignorance of the fate of the great Arctic expedition of 1845, the most startling intelligence reached us—intelligence that reopened the wound in the national heart which time was beginning to heal. It suddenly became known, late in autumn 1854, that Dr. John Rae, chief factor in the employment of the Hudson's Bay Company, and one of the most intrepid of British explorers, had discovered and brought home information and relics which were conclusive as to the unspeakably mournful fate of at least one-third of Franklin's officers and men.

This bold and successful explorer, certainly the most successful of all Arctic travellers engaged in the Franklin search—with perhaps the exception of McClintock, who only followed up the path which Rae pointed out, though he arrived at more important results than the original discoverer—has had a singularly adventurous, useful, and interesting career. With the exception of one modest volume, in which he records the results of his expedition from Repulse Bay across Rae Isthmus, and along the southern shores of Boothia Gulf, Dr. Rae may be said to have published nothing. Yet there are few modern travellers whose boat and sledge achievements, apart from his famous expedition of 1853–54, in which he ascertained the fate of Franklin and his companions, and found many important relics of the party, are more worthy of public attention. Fortunately the present writer has been enabled to place before readers the first published narrative of the career and principal exploits of this explorer, from original notes, journals, etc., kindly supplied by Dr. Rae himself for the purposes of the present work.

The Orkney Islands, a small group off the northern land's end of Scotland, has sent out a greater number of hardy and capable navigators to the Arctic seas than any other district of equal area in the British dominions. On one of these John Rae was born, at about the time when the great victory of Waterloo brought a long period of confusion and alarm to a close, and enabled men once more to turn their attention in security toward the pursuits of peace. Before the age of boyhood had passed, he was studying medicine at Edinburgh University; but in the meantime he had already received an education of another sort on the coasts and amid the barren moors and hills of the stormy Orkneys. "I there acquired as perfect a knowledge of boating," writes Rae, "as could be obtained by constant practice; because to my brothers and myself our boat was our chief plaything. In it we used to put to sea in all weathers, the stormier the better, and we stayed out as long as it was possible to remain at sea in any small undecked craft. Our father had given us a beautiful, fast-sailing boat, of about eighteen feet, that could beat anything of her size in that part of the world. We got the boat and also her small tender for fishing, on condition that we kept them in good order; and this we did very effectually, as we took great pride in having every rope and all her four sails in perfect trim. We lived opposite the stormy 'Hoymouth,' and were exposed to constant gales from the west, which brought very heavy waves direct from the Atlantic; so we had abundant opportunities of learning boatmanship under the most trying of situations—fighting against heavy seas and strong currents running at from six to eight miles an hour. I mention this experience only because it stood me in good stead afterwards, in my Arctic work. I also learned to shoot as soon as I was old enough to lift a gun to my shoulder."

Rae passed as surgeon in Edinburgh in 1833, before he was twenty years of age, which seems to show that the splendid physical training of his youth interfered in no way with the due cultivation of his mind. In the same year (1833), he went out as surgeon on board one of the

Hudson's Bay Company's ships. "On the way home," he continues, in his *naive*, unassuming, downright fashion, "we were stopped by an impenetrable barrier of ice in Hudson Strait. At that time the Company presented a *bonus* to every captain who brought his ship home to England, and, stimulated by the expectation of this gratuity, our chief officer exerted every effort to force a passage. All in vain, however. We were obliged to turn back. New ice was now forming all round us, and so much of it clung to the forepart of the vessel (it was about two feet thick on the forecastle), that the extra weight brought her down three feet by the head.

"We went to an island called Charlton, in James Bay—covered with snow when we reached it—and found some old houses, which we repaired for winter quarters. The ship was laid on shore under shelter of a point, and the cargo taken out and placed under a tent extemporised from the sails. We had scarcely any fresh meat with us, and little lime juice or vegetables; so it was not surprising that scurvy attacked the party. Of the seventeen persons attacked, two died before the spring, and some of the others were in a very dangerous condition, when, fortunately the spring sun cleared the snow off the ground, and we found abundance of cranberries (a famous anti-scorbutic). The sick men were taken out during the warm part of the day, and left to eat as many berries as they wished. These, with some soup, made at a later date from the bud of the vetch, restored the health of the invalids without almost any other anti-scorbutic; for the small quantity of fresh meat obtained did not amount to over a few days' rations per man. My first survey work was to examine the whole shore of our island in a bark canoe, as soon as the ice cleared away a little. This work occupied us—I had two sailors with me—three days.

"Thinking from what I saw that I should like the wild sort of life to be found in the Hudson's Bay Company's service, I accepted the appointment of surgeon at Moose Factory, the former medical man being about to resign his situation. It was at this place, on the southwest shore of James Bay—the southern arm of Hudson's Bay—that I learned all the different modes of hunting, fishing, sledge-hauling, snowshoe-walking, and camping out, both in winter and summer, spring and autumn, that were afterwards so useful to me in my Arctic expeditions.

"Some of my adventures on my hunting excursions were curious enough, and occasionally dangerous. One night I and a young friend were encamped on a low flat island some miles out to sea, at the mouth of the river. What I have dignified by the name of encampment was the small birch-built canoe turned up to windward, a bit of oil-cloth under us to keep us out of the mud, a couple of blankets as bedding, and a fold of the oil-cloth over us to keep off the rain. The night was a pitch-dark one, when a gale of wind came on from seaward, which brought the tide upon us. We quickly righted the canoe, and bundled our things into her. But where to go was the question! To attempt to reach the main shore, if we had even known exactly where it was, would have been futile, as we would have been filled in a moment among the rough waves roaring not far from us. On a neighbouring flat island, separated from that we were on by a very narrow channel not more than one hundred yards wide, there was a small 'clump' of willows six or seven feet high. To reach this clump, about a mile off, was our only chance of safety; and I gradually pushed the canoe as the tide rose higher and higher, across from our island, in the direction, as I thought, of this willow haven. It was so dark that I could not see the bow of our own canoe, but by narrowly watching the effect of the rising water, I kept a fairly good course. After about an hour of this work, all at once I could not touch bottom with my paddle, and felt the canoe was in a current running like a mill stream. This I knew was the channel between the two islands, but was I right for the willows which only occupied some fifty square yards? We paddled vigorously across, and I have had few more pleasurable sensations than when I felt—I could not see them—the bow of the canoe scrape against some branches, and we were safe. The canoe was pushed in some way among the bushes and fastened, stern to windward, under their lee. A snowstorm came on, and continued all night; a roaring sea was raging within fifteen yards of us; and my dear companion—now, alas, dead and gone—not being so habituated as myself to this work, shook so with cold that the tremor was communicated to our canoe, and his teeth chattered so as to be quite audible. The tide rose until only a foot and a half of our willows remained above water; and by the morning's light a number of inches of snow had fallen. The tide at last fell enough to allow me to walk about, and after making my friend as comfortable as I could, I commenced shooting snow geese which were crowding about the land just left exposed. Having killed a number I went back to the canoe, but could scarcely find my companion, the snow having fallen so thick as to cover him up. Fortunately a piece of wood had got entangled among the willows. This was cut up for a fire, and a cup of tea made us comfortable. All this time I may mention that I was very wet, having no waterproof coat on, but I had no impression of being particularly uncomfortable.

"It came on so bitterly cold, that for fear of injury to my companion I paddled some nine miles against a strong current, my friend being quite helpless, and although my wet clothes and moccasins froze hard, I suffered no bad effects. I could multiply such like events by the half dozen, but I merely mention this one to show one instance of the sort of training, if it may be so called, that I went through to educate me for Arctic service. In winter, snowshoe-walking and sledging were among my chief amusements, so, after a ten years' life at Moose Factory, I had learned a good many useful lessons as to the best manner of taking care of myself in cases of difficulty.

"Sir George Simpson had the kindness to offer me the command of the Arctic expedition of which I took charge in 1846. I had the choice whether to take a fine small schooner or small open boats; I chose the latter, because they could work between the ice and the shore, and in the event of a difficulty could be hauled upon the floe.

"The work to be done, namely, the joining of the surveys of Parry and Ross, had baffled Parry himself, as well as a second Government expedition under Lyon, and a third in the 'Terror,' under Back. I went and accomplished the work at an expense of about £3000, and wintered in a manner never before or since attempted except by myself and my gallant fellows. Hall the American did something of the same kind, but with this difference, that he was landed by a whaling ship close to Repulse Bay, and he had always within reach one or more whalers upon which he could have retreated if necessary. My nearest aid was six hundred or eight hundred miles off."

The expedition which Dr. Rae here alludes to has already been described. It need only be added here that the amount of money spent by the British Government in the three expeditions of Parry, Lyon, and Back, all of which failed in carrying out the proposed survey, could not have been less than £100,000.

Having successfully concluded the expedition of 1846–47, and examined and surveyed the southern shores of Boothia Gulf, from Parry's farthest (Fury and Hecla Strait) on the east side, round to Ross's farthest (Victoria Harbour) on the west side, Rae returned to England in the autumn of 1847. At that time Sir John Richardson was preparing to set out on an expedition along the shores of the Polar Sea, in search of his friend and former comrade, Franklin; "and," says Rae, "he asked me to accompany him. This offer was very complimentary, for Sir John had received hundreds of applications from men in almost all ranks and stations to be allowed to go with him as second. After due consideration, I accepted the offer, and went with the boats from the Mackenzie, along the coast to the Coppermine, near which the ice blocked the way, and we had to abandon the boats before we reached the river (see page 468). Again in 1849 (see page 480) I visited the Arctic Sea *via* the Coppermine, but the ice was so closely packed in the direction I was told to take, that we could make no headway; and on returning to the Coppermine, and attempting to ascend it, our boat, owing to the mismanagement of the steersman, was lost, and our excellent Eskimo interpreter drowned. After this I remained one winter in charge of the Mackenzie River district; and in 1850 was again employed on Arctic service."

In the spring of 1850, whilst Rae was the officer in charge of the large Mackenzie River district for the Hudson's Bay Company, he received a despatch from the Governor of the Company, Sir George Simpson, informing him that Her Majesty's Government had asked for the "loan of his services" to command a boat expedition, to follow up the search for Franklin. The intimation expressed high appreciation of Dr. Rae's abilities as an explorer, and left him in every respect untrammelled. He was to select whatever route he thought most promising, and was to conduct the enterprise in every way as he thought best. "I had no other instructions of any kind," says Rae; "but I was placed in a most difficult position, because a boat voyage, under a naval officer (Commander Pullen), had already searched the Arctic shores from Point Barrow to the Mackenzie; and the same officer was this season (1850) to examine the coast eastward from the Mackenzie. The only line of route left, the examination of which was still unprovided for, was that lying eastward from the Coppermine, but we had no small boats such as were absolutely requisite for this route, and to build them at Fort Chipewyan, where Simpson's had been built, or even at Fort Simpson, would have prevented us getting to the coast earlier than 1852. Simpson has said in his very excellent narrative that there was no wood at the north-east end of Bear Lake fit to build boats of. Notwithstanding this, I determined to make the attempt.

"We went in the autumn of 1850, with two large boats, very scantily supplied with provisions, to Fort Confidence, and immediately commenced operations. After a careful search, a clump of moderately good trees was found, which the carpenter thought could be cut into planks suitable for boats. A sufficient number of these were cut down, and boated to the fort in a wonderfully short space of time, for we had to hurry forward, as the winter might come on any day, and shut up navigation. A difficulty, however, occurred; our carpenter, a very good one, could build the large river boats very well, because he had models to go by, but he had never seen boats such as I required. I therefore had to draft them, so that the width, shape, etc., of every plank could be measured before being put on the frame-work. In this way two very fine little boats of about 22 feet keel and 7 feet beam were constructed under great difficulties; for, as will be evident, the wood had no time to season. But it will be seen by what follows, they did their work admirably. Another difficulty was the making of the sails. These, after I had cut them out, were sewn by some of my people, but roped chiefly by myself in all important parts; the rigging being all fitted and spliced with my own hands.

"We spent a very cold winter, frequently on reduced rations, for the Indians could not bring us as much food as we required, and the quantity we had brought with us was, as I have already said, very small; yet we had enough to keep us in good health, although with no vegetables of any kind, very little flour and tea, and *no rum*, which I have never used on Arctic service, believing its use to be most prejudicial."

Rae's search expedition of 1851 consisted of a rapid but effective sledge journey from Great Bear Lake to Wollaston Land in the spring, and a summer exploring excursion along the south and east coast of Victoria Land in boats. In order to maintain the continuity of our narrative, it will be necessary briefly to summarise the reports of these journeys, and to note their interesting results.

Sledge Expedition, 1851.—Dr. Rae left Fort Confidence on the 25th April, arrived at Provision Station on Kendall River on the 27th, and thence made his actual start for the shore on the 30th, with two men and two sledges drawn by dogs. A fatigue party of three men and two dogs accompanied him to within ten miles of the coast, and he was thus enabled to take forward depots of provisions for the return journey. On the 2d May he reached Richardson Bay, about five miles west of the mouth of the Coppermine. Having resolved to travel by night, to avoid the sun-glare, Rae started with his two companions at ten P.M. on the 2d. Travelling along over the ice in an east-north-east direction, and passing Point Lockyer and Cape Krusenstern, he crossed the frozen Dolphin and Union Strait, and arrived on Douglas Island at three A.M. on the 5th. Starting again the next evening, he walked across the narrow strait between Douglas Island and Wollaston Land. He was now on entirely new ground, and, turning eastward, he marched along, examining the shore, which was found in the main uninteresting, and affording no traces of having ever been visited by Europeans. "We built snow-huts every night when cold enough to require them," says Rae, "and all our bedding, for three persons, amounted to about 15 or 16 lbs., consisting of one blanket and a half, and three narrow strips of hairy deerskin to lie upon. The heat of our bodies did not thaw the snow." On the 7th the snow-hut was erected in lat. 68° 31', long. 111° 30' W., under a steep bank, surmounted by some whitish limestone and reddish-brown sandstone *in situ*. In all his explorations, Rae has always wonderful luck with the rifle, and at this spot he shot no less than ten hares during the interval between taking the observations for time and latitude. "These fine animals were very large and tame," he writes, "and several more might have been killed, as well as many partridges, had I thought it expedient to follow them."

Pushing on eastward, Rae discovered and named the Richardson Islands and Welbank Bay. As he travelled onward on the night of the 9th, the land continued low, and had an easterly trending. The thermometer showed a temperature of 22° below zero, as some protection against which the shelter of the snow-hut was more than usually acceptable. On this night one of the men was somewhat deeply frost-bitten in the face, and Rae found that taking a set of lunar distances was rather chilly work. "I have generally found, indeed," he remarks, alluding to the degree of cold experienced on the night of the 9th May, "that a temperature which in winter would be pleasant, is in the latter part of spring almost insupportably cold. The latitude of our position was 68° 37' 48" by observation; longitude, by account, 110° 2'."

From this spot the farthest point of land bore east-south-east, but Rae did not think it necessary to advance farther eastward along this coast, because his survey and that of Dease and Simpson met at this point; and had he gone on farther east, he would have been going over ground already discovered and roughly surveyed. The object of Rae's search along the south coast of Wollaston Land was to seek for some strait that, leading northward, might afford a passage in the direction of the region in which it was believed by many the "Erebus" and "Terror" were to be found. No such strait was found eastward from the meridian of Douglas Island to long. 110° 2', beyond which, in an easterly direction, the land had been examined by Dease and Simpson to Cambridge

Bay. It would therefore be unavailing to travel on toward the east in search of a northward running strait. Besides, Rae had only a few more days to spare. His two boats were to be ready for him by the middle of June, and it was expected that he should start on his boat expedition not later than the middle of the month. His time was therefore strictly limited. "There were now," he says, "two courses open to me—the one, to strike overland to the north, in search of the sea-coast; the other, to return along the coast and travel westward, in hopes that some of the spaces of Wollaston Land, left blank in the charts, might prove to be the desired strait. I chose the latter of these courses." The journey westward to Douglas Island, where the party arrived at eight A.M. on the 15th May, was favourable. Thence Rae proceeded west along the shore of Wollaston Land. Simpson's Bay and the Colville Hills were successively discovered and named, and on the morning of the 22d Cape Hamilton, a limestone cliff at least 170 feet high, was reached. "A couple of miles to seaward," says Rae, "there were thirteen Eskimo lodges, and we had an amicable interview with the poor harmless inhabitants, who were rather timid at first, but soon gained confidence. It was difficult to make them understand that no return was expected for some presents I made them. None of the women showed themselves, but all the men were well and cleanly dressed in deerskin. They were all very fat, having evidently abundance of seals' flesh and fat, large quantities of which were carefully deposited in sealskin bags under the snow. We purchased a quantity of this for our dogs, and some boots, shoes, and sealskins for our own use. After a most friendly interchange of signs and words, few of which could be understood on either side, we parted." Pullen Point and Lady Richardson Bay were discovered and named.

On the night of the 22d the coast which had hitherto had a northwest trend bent round toward the north-east. The day's journey ended on the shores of a small bay in lat. 70°, and long. 117° 16'. "The period I had allowed for our outward journey having now arrived," writes Rae, "I left our dogs and one of the men here; whilst with the other man I travelled half a day's journey farther. At 8.30 on the night of the 23rd the night was beautiful; and as we started with no other encumbrance than a gun, telescope, and compass, we travelled fast over the hard snow and ice. After walking two miles to the north-west we turned a cape, which received the name of Baring (in honour of the First Lord of the Admiralty), beyond which the coast took a sudden bend to east by north for eight miles, and then became more northerly for six and a half miles, which was the farthest point reached. . . . Near the place from which I turned back the land was fully three hundred feet high, from which objects could be seen at a great distance; and some land fifteen or twenty miles off was observed, the most westerly point bearing N. 25° W. . . . It is difficult," continues Rae, "to determine whether the water dividing these two shores is a bay or a strait, but from the little information I could obtain from the Eskimos I suspect it to be the latter." Captain Collinson, in the "Enterprise," however, explored this "dividing water" in the summer of 1852, and found it a deep inlet surrounded by land. It now appears on the map as Prince Albert Sound.

On the 24th May Rae commenced his homeward journey, checking his bearings, distances, observations for latitude, etc., as he proceeded. He describes all the land from Cape Baring on the south side of the entrance to Prince Albert Sound to Cape Lady Franklin, opposite Douglas Island, as being extremely barren, and destitute of herbage sufficient to tempt the deer migrating northward from the mainland to pause in their journey into the interior of Wollaston Land. On the 2d June he had reached Cape Hearne; on the 4th he encamped on Richardson Bay; and on the appointed day, the 10th June, he arrived at his starting-point, Provision Station, Kendall River, "having," he says, "been five days coming from the coast, during some of which we were fourteen hours on foot, and continually wading through ice-cold water or wet snow. . . . Our principal food," he continues, "was geese, partridges, and lemmings. The last being very fat and large, were very fine when roasted before the fire or between two stones. These little animals were migrating northward, and were so numerous that our dogs, as they trotted on, killed as many as supported them without any other food."

This sledge journey, extending to eleven hundred miles, including the distance from Fort Confidence to Kendall River, was the fastest on record—averaging twenty-three miles a day—or, not counting three days on which Rae and his men were compelled to keep inside their snow-hut owing to bad weather, twenty-four and a half miles a day.

Boat Expedition, 1851.—Exactly three days after Rae's return from travelling by sledge along the south shores of Wollaston Land, his boats from Fort Confidence, where they were built, were brought to the rendezvous, Provision Station, Kendall River, and everything being in readiness, the explorer and his party started on the boat expedition toward the south and east coasts of Victoria Land two days after, namely, on the 15th June 1851. While descending the Coppermine, which was much swollen, six deer and four musk-oxen were shot on the 23rd and 24th, and the greater part of their flesh partially dried over a fire for future use. At the close of the month forty salmon and white fish were taken in a net at Bloody Fall in fifteen minutes, and at the mouth of the river they killed deer, fish, and geese in abundance. Throughout the entire voyage game were abundant, but as the party were plentifully supplied with provisions, Rae did not think it worth while to follow them. Going on the same principle, we shall not think it necessary to return to this subject again in the course of our necessarily brief notice of the boat expedition of 1851 to Victoria Land.

Coasting eastward from the mouth of the Coppermine along the north shore of America, through a narrow channel between the sea-ice and the beach, Rae rounded Cape Barrow on the 16th July, reached Cape Flinders on the 22d, and Cape Alexander on the 24th. The ice in Dease Strait between Victoria Land and the American mainland remained unbroken until the 27th, when Rae pushed his way across among the loose pieces to the Finlayson Islands, and thence to the mainland on the west side of Cambridge Bay. On the 1st August the party reached Cape Colborne, the most easterly point on this coast examined by Dease and Simpson. All the coast of Victoria Land *east* from this point *was new*; and Rae entered upon the examination and survey of it with his usual zest. At Cape Colborne the shores of Victoria Land are high and steep, but toward the east they are considerably lower. Anderson, Parker, and Stromness Bays, and Macready and Kean Points were successively discovered and named. The coast was found to trend to the north-east. On the 3rd August, after making a successful run of one hundred miles without stopping except to cook, the party reached lat. 69° 12', long. 101° 58'. "On the 4th," says Rae, "the wind again set in from the north, increasing to a perfect gale; and although we could gain ground pretty fast by plying to windward, our slightly-built craft strained so much in the heavy seas that frequently washed over us—in fact, one of the boats had a plank split—that we lowered sails on gaining a partial shelter from the land, and after a tough pull of two miles, during which we were sometimes barely able to hold our ground, we entered a snug cove and secured our boats." Prince Albert Edward Bay was discovered on the 9th, but as there were no evidences that Eskimos had recently visited its shores, and no signs that Europeans had ever been on the coast, Rae pushed on northward without pausing to examine it and reached lat. 69° 42'. At this point north-easterly winds put a stop to all farther progress.

"On the 12th," writes Rae, "finding that there was little or no prospect of change in the wind, preparations were made for a foot journey of a week's duration to the northward. Leaving, therefore, directions that one of the boats should follow us along shore if the ice cleared away, I started a short time before noon, in company with three men; and as we trusted to killing both deer and geese on our way, we carried with us provisions for only four days. Hoping to avoid the sharp and rugged limestone *débris* with which the coast was lined, we at first kept some miles inland, but with trifling advantage, as the country was intersected with lakes, which obliged us to make long detours. Nor was the ground much more favourable for travelling than that nearer the beach, being, in fact, as bad as it could be, in proof of which I may mention that in two hours a pair of new moccasins, with thick, undressed buffalo-skin soles, and stout duffel socks, were completely worn out; and before the day's journey was half done, every step I took was marked with blood. We gained a direct distance of seventeen miles after a walk of twenty-four hours, and bivouacked near the shore.

Although we had passed a good many fine pieces of driftwood some time before, here we had some difficulty in collecting enough to boil the kettle. Opposite our resting-place, and not far from shore, was an island some miles in extent, to which I gave the name of Halkett. Next morning, when we had travelled three miles northward, a large piece of wood was found, very opportunely, about breakfast-time. As the travelling continued as bad as ever, and as the whole party were more or less footsore, I resolved to remain here, to obtain observations, during which time two of my men pushed on ten miles to the north, and the other went to kill deer." The results of the observations here were lat. 70° 2', long. 101° 24'. In the evening the two men returned after their ten miles' walk northward. From their farthest point they could see to a distance northward of seven miles; and to this farthest land Rae gave the name of Pelly Point.

Here, then, Rae's discoveries in 1851 ended, and from this point the return journey commenced on the following morning, the 13th August. But it must have been with some little degree of complacency that, standing on the dreary, flat, and stony shore, he looked around upon coast and frozen strait that had never before been surveyed by civilised man—unless, perhaps, Franklin's crews, or a party belonging to his expedition, had wandered hither, after breaking out from their winter quarters at Beechey Island in 1845–46. His boat expedition along the American shore and the south and east coasts of Victoria Land, was the longest but one—that of Dease and Simpson—ever made in this region. "I discovered and named Victoria Channel, down which the Franklin ships were driven," writes Rae, "and reached, with my boats, coming from the south, a latitude higher than that in which the 'Erebus' and 'Terror' were abandoned. I knew that Victoria Strait was not a bay, because the flood-tide came from the north." On the 15th he took possession of his discoveries in the name of her Majesty. From this date onward the homeward journey was prosecuted prosperously.

On the 20th a piece of pine wood was found, resembling the butt-end of a small flag-staff. A piece of white rope was fastened to it, in the form of a loop, by two copper tacks. Both the rope and the tacks bore the Government mark, the broad arrow being stamped on the latter, and the former having a red worsted thread running through it. Half-a-mile farther on, a piece of oak, 3 feet 8 inches long, and the one-half of which was squared, was picked up. Rae regarded it as a boat's stanchion. Writing in 1851 respecting these pieces of wood, which, without doubt, had been worked with European tools, Rae says: "As there may be some difference of opinion regarding the direction from which those pieces of wood came, it may not be out of place to express here my own opinion on the subject. From the circumstance of the *flood-tide coming from the northward, along the east shore of Victoria Land,* there can be no doubt that there is a water-channel dividing Victoria Land from North Somerset, and through this channel, I believe, these pieces of wood have been carried, along with the immense quantities of ice that a long continuance of northerly and north-easterly winds, aided by the flood-tide, had driven southward." Subsequent discoveries proved that these fragments came from the lost Franklin ships. On the morning of the 24th the breeze that had been blowing from south east by east gradually increased to a gale. Reef after reef was taken in, until the two small boats were scudding under the smallest canvas. A very heavy sea was running, which broke over, now and then, from stem to stern, and bent and twisted the slight-built but fine little craft in every direction. "At last," writes Rae, "the weather became so bad that I was reluctantly obliged to look out a harbour. This was dangerous work, as we had to run almost among the breakers before it was possible to see whether the place we made for would afford a shelter. But we were fortunate; and at 9.30 A.M., when eight miles north-east of Cape Peel, we were snugly moored in a small land-locked bay, the entrance into which was not twenty yards wide."

Point Ross was reached on the 28th, and from this there was an uninterrupted run to the mouth of the Coppermine. After five days of arduous labour, the boats were dragged up the swollen river. On the 5th and 6th the party ascended the Kendall, and on the afternoon of the 10th they arrived at Fort Confidence, at the eastern extremity of Great Bear Lake.

The results of the sledge and boat expedition conducted by Rae in 1851 were the discovery and survey of 725 miles of previously unknown coast-line of the shores of Wollaston and Victoria Lands. For his discoveries in this expedition, the Royal Geographical Society awarded their highest honour—the founder's gold medal—to Dr. Rae.

After reaching Bear Lake, Rae started with his men to travel to Red River (now Winnipeg) Fort, one of the stations of the Hudson's Bay Company, near the northern frontier of Dakota, U.S. Referring to this feat, Rae says: "On our homeward journey from Bear Lake to Red River we made a forced march on snow-shoes of more than 1300 miles, chiefly to save the expense to Government of five or six months' wages for the men. From Red River I went to Minnesota, a distance of 450 miles, which was accomplished in ten days, at the rate of 50 miles a day—one day being spent under shelter from the weather. In fact, from starting on the 25th April 1851, on our sledge journey to the coast, till the spring of 1852, I and my party were continually on the move either over ice, in boats, or on snowshoes." The entire distance travelled between the dates named was over 8000 miles.

CHAPTER TWO

Rae's Last Arctic Expedition—Extraordinary Intelligence—List Of Relics Found By Rae—Winter At Repulse Bay—Conclusion.

In the summer of 1853 Dr. Rae was appointed by the Hudson's Bay Company to the command of an expedition planned and proposed by himself, organised mainly for geographical purposes, and also, no doubt, for the purpose of forwarding the interests of the Company, the two objects being in a sense identical. We have already seen that Dr. Rae was employed by the Company on a similar expedition in 1846–47. He was then asked to aid in completing the survey of the Arctic shores of British North America—the special duty marked out for him being to penetrate across the unknown land from Repulse Bay to the southernmost arm of Boothia Gulf; to survey the shores of that gulf northward, so as to connect his discoveries with those of Sir John Ross; and to ascertain, beyond doubt, whether any navigable passage led westward from the gulf into the Arctic Sea on the west of Boothia Peninsula. It may be remembered that he conducted this expedition with great capacity and spirit to a most satisfactory termination, and ascertained that no waterway led west from the gulf into the Arctic Sea to the south of the parallel of about 70° N., down to which point this coast had previously been surveyed by Sir John Ross. He was now called upon to continue the work thus auspiciously begun—his orders in 1853 being to complete the survey of the *west* coast of Boothia, as in 1846 they had been to survey the east coast of that great peninsula.

Dr. Rae's genius is eminently practical, and he goes to his point without deviation or delay. His labours were again to commence (as in 1846) at Repulse Bay. He therefore sailed from the north of Hudson's Bay, north through Rowe's Welcome to Repulse Bay, which he reached on the 14th August 1853. The spot at which he landed was about seven miles east of his old winter quarters in 1846–47, but on the following day he sailed down to near the old familiar locality at the mouth of North Pole River. Mooring his boat here, he landed and pitched his tents. The weather was dark and gloomy, "and," says Rae, "the surrounding country presented a most dreary aspect. Thick masses of ice clung to the shore, whilst immense drifts of snow filled each ravine and lined every steep bank that had southerly exposure. No Eskimos were to be seen, nor any recent traces of them. Appearances could not be less promising for wintering safely, yet I determined to remain until the 1st September, by which date some opinion could be formed as to the practicability of procuring sufficient food and fuel for our support during the winter; all the provisions on board at this time being equal to only three months' consumption. The weather fortunately improved, and not a moment was lost. Nets were set, hunters were sent out to procure venison, and the majority of the party was constantly employed collecting fuel. By the end of August a supply of the latter essential article (*Andromeda tetragona*) for fourteen weeks was laid up; thirteen deer and one musk-bull had been shot, and one hundred and thirty-six salmon caught." This was a fair business-like commencement towards accumulating food and fuel for the winter, and it may be remarked here that it seems to be a maxim with Dr. Rae that a country should always feed its explorer. The entire absence of Eskimos from their customary haunts in the neighbourhood caused him considerable anxiety; not that he expected any aid from them, but because he could only interpret their absence as proof that the locality no longer yielded a plentiful supply of venison, owing most probably to the circumstance that the deer had ceased to pass this locality in their migrations to and from the north.

On the 1st September it was necessary to decide whether to stay here or return, and as the Doctor did not wish to conceal from his men the risk of being frozen in on this apparently desolate and barren shore, he called them together, informed them of the slender store of provisions in hand, and remarked on the unflattering prospect of obtaining sufficient supplies during the winter months. But the men knew Dr. Rae. All of them volunteered to remain. "Our preparations for a nine months' winter," writes Rae, "were continued with unabated energy. The weather, generally speaking, was favourable, and our exertions were so successful, that by the end of the month we had a quantity of provisions and fuel collected adequate to our wants up to the period of the spring migrations of the deer. One hundred and nine deer, one musk-ox (including those killed in August), fifty-three brace of ptarmigan, and one seal, had been shot, and the nets produced fifty-four salmon. Of larger animals above enumerated, forty-nine deer and the musk-ox were shot by myself, twenty-one deer by Mistegan, the (Indian) deer-hunter, fourteen by another of the men, nine by William Ouligbuck (Eskimo interpreter), and sixteen by the remaining four men." From the above it appears that Rae's party consisted of seven persons besides himself.

During September and October the party lived in tents; but at the close of the latter month the cold became very severe, and the snow freezing hard, Rae was able to build snow houses, which afforded palatial accommodation and comfort compared with the tents. Few deer were shot during the winter, and fish were caught in inconsiderable quantities. On two occasions, on the 1st and the 27th February, a singular phenomenon was witnessed. Rae describes it as "that beautiful but rare appearance of the clouds near the sun, with three fringes of pink and green following the outline of the cloud." This splendid phenomenon was often seen during the spring, and was usually followed by a day or two of fine weather.

Having set up a carpenter's workshop built of snow, and constructed a number of sledges to be used in the spring journeys, Rae set out on the 14th March with three men, dragging sledges with provisions to be placed *en cache* in advance. The party pushed on as far as Cape Lady Pelly, on the west shore of the extreme south of Boothia Gulf. Here the provisions were deposited under a heap of huge stones, secure from all marauders except bears and men. From this point Rae returned, and arrived at Repulse Bay on the 24th, having walked altogether 170 miles in ten days. On the 31st March the great spring journey was commenced, Dr. Rae taking with him four men, including the interpreter Ouligbuck, and an amount of provisions, which, taken together with the quantity deposited at Lady Pelly Bay, would be sufficient for sixty-five days. The object of the journey was to cross Boothia Peninsula from Pelly Bay to the Castor and Pollux River, discovered by Simpson, and thence to survey the west coast of Boothia northward to Bellot Strait, and thus connect Simpson's discoveries with those of Kennedy and his lieutenant, the gallant Bellot.

On the 6th April the party arrived at their depot on Cape Lady Pelly, from which they took up their provisions. On the 10th they reached Colville Bay, on the west shore of Committee Bay, and in latitude about 68° N. On the morning of the 17th Rae reached the shore of Pelly Bay, in making a troublesome but unavoidable detour across which three days were spent. Fresh footmarks of an Eskimo and the track of a sledge were observed on the 20th, and Rae sent his interpreter and one companion to look for natives. After an absence of eleven hours the men returned, bringing with them seventeen Eskimos (five of them women). "They would give us," says Rae, "no information on which any reliance could be placed, and none of them would consent to accompany us for a day or two, although I promised to reward them liberally. *Apparently there was a great objection to our travelling across the country in a westerly direction.** Finding it was their object to puzzle the interpreter and mislead us, I declined purchasing more than a piece of seal from them, and sent them away." On the 21st the party started westward across the peninsula. They had not proceeded far, when they were met by a very intelligent Eskimo driving a dog-sledge laden with musk-ox beef. This man readily consented to accompany Rae two days' journey. He explained that the road by which he had come would be the best for the party. Shortly after this the party was joined by another

---

* "I found that it was their favourite hunting ground for musk-oxen, deer, etc., and that the natives had *caches* of provisions in that direction." —Dr. J. Rae.

Eskimo, who had heard of white men being in the neighbourhood, and was curious to see them. Here we must quote somewhat freely from Rae's brief narrative: "This man (the new-comer) was very communicative; and on putting to him the usual questions as to his having seen 'white men' before, or any ships or boats, he replied in the negative, but said that *a party of 'Kabloonans'* (whites) *had died of starvation a long distance to the west of where we then were, and beyond a large river.* He stated that he did not know the exact place, that he never had been there, and that he could not accompany us so far. The substance of the information then and subsequently obtained from various sources," continues Dr. Rae, "was to the following effect:

"In the spring four winters past (1850), whilst some Eskimo families were killing seals near the north shore of a large island, named in Arrowsmith's charts, King William Land, forty white men were seen travelling in company southward over the ice, and dragging a boat and sledges with them. They were passing along the shore of the above named island. None of the party could speak the Eskimo language so well as to be understood; but by signs the natives were led to believe *the ship or ships had been crushed by ice*, and that they were then going to where they expected to find deer to shoot. From the appearance of the men (all of whom, with the exception of one officer, were hauling on the drag-ropes of the sledges, and were looking thin), they were then supposed to be getting short of provisions, and they purchased a small seal, or piece of seal, from the natives. The officer was described as being a tall, stout, middle-aged man. When their day's journey terminated, they pitched tents to rest in.

"At a later date the same season, but previous to the disruption of the ice, the corpses of some thirty persons and some graves were discovered *on the continent*, and five dead bodies on an island near it, about a long day's journey to the north-west of the mouth of a large stream, which can be no other than Back's Great Fish River, as its description and that of the low shore in the neighbourhood of Point Ogle and Montreal Island agree exactly with that of Sir George Back. Some of the bodies were in a tent or tents, others were under the boat, which had been turned over to form a shelter, and some lay scattered about in different directions. Of those seen on the island, it was supposed that one was that of an officer (chief), as he had a telescope strapped over his shoulders, and his double-barrelled gun lay underneath him. *From the mutilated state of many of the bodies,* and *the contents of the kettles,* it is evident that our wretched countrymen had been given to the last dread alternative—cannibalism—as a means of sustaining life. A few of the unfortunate men must have survived until the arrival of the wild-fowl (say until the end of May), as shots were heard, and fresh bones and feathers of geese were noticed near the scene of the sad event.

"There appears to have been an abundant store of ammunition, as the gunpowder was emptied by the natives in a heap on the ground, out of the kegs or cases containing it, and a quantity of shot and ball was found below high-water mark, having probably been left on the ice close to the beach, before the spring thaw commenced. There must have been a number of telescopes, guns (some of them double-barrelled), watches, compasses, etc., all of which seem to have been broken up, as I saw pieces of these different articles with the natives; and I purchased as many as possible, *together with some silver spoons and forks,* an order of merit in the form of a star, and a small plate engraved 'Sir John Franklin, K.C.B.'"

The following is Dr. Rae's list of the articles belonging to the officers of the "Erebus" and "Terror," which he purchased from the Eskimos of Boothia, in 1853–54, viz.: One silver fork—crest, an animal's head with wings extended above; three silver forks—crest, a bird with wings extended; one silver table-spoon—crest, with initials, "F.R.M.C." (Captain Crozier, "Terror"); one silver spoon and one fork—crest, bird with laurel branch in mouth, motto, *Spero meliora*; one silver table-spoon, one tea-spoon, and one dessert-fork—crest, a fish's head looking upwards, with laurel branches on each side; one silver table-fork—initials, "H.D.S.G." (Henry D. S. Goodsir, assistant-surgeon, "Erebus"); one silver table-fork—initials, "A. McD." (Alexander McDonald, assistant-surgeon, "Terror"); one silver table-fork—initials, "G.A.M." (Gillies A. McBean, second master, "Terror"); one silver table-fork—initials, "J.T.;" one silver dessert-spoon—initials, "J.S.P."

(John S. Peddie, surgeon, "Erebus"); one round silver plate, engraved "Sir John Franklin, K.C.B;" and a star or order of merit, with motto, "*Nec aspera terrent*, G.R. III., MDCCCXV."

None of the Eskimos with whom Rae came in contact had ever seen the "white men," either before or after death, nor had they ever been at the place where the corpses were found, but had obtained their information from natives who had been there, and who had seen the troop of starving mariners travelling over the ice.

The foregoing narrative of the results of Dr. Rae's interviews with the Eskimos of Boothia, is extracted from the published account of his expedition, which the explorer wrote to the Directors of the Hudson's Bay Company. This letter, dated from York Factory, Hudson's Bay, September 1st, 1854, on the day after his arrival from Repulse Bay, was necessarily hurried and imperfect. Further particulars afterwards suggested themselves, but have never yet been published. It is with great pleasure, therefore, that the present writer is enabled to present to the public the following notes, embracing fresh particulars in connection with this most interesting episode in Arctic Exploration. These valuable notes have, in the kindest manner, been supplied by Dr. Rae for the present work. We give Dr. Rae's communication, so courteously sent, in the form in which it has come to hand:

"When travelling westward on my spring journey, I met an Eskimo, to whom we put the usual question, 'Have you seen white men before?' He said, 'No, but he had heard of a number having died far to the west,' pointing in that direction. Noticing a gold cap-band round his head, I asked him where he obtained it, and he said it had been got where the dead white men were, but that he himself had never been there, that he did not know the place, and could not go so far, giving me the idea that it was a great way off. I bought the cap-band from him, and told him that if he or his companions had any other things, to bring them to our winter quarters at Repulse Bay, where they would receive good prices for them. Some further details were obtained on our way home, and the purchase of one or two additional articles was effected; but it was not until our arrival at Repulse Bay, that I could gain information as to the locality where our countrymen had perished—for I clearly made out that they must have all died some years before, or they must have reached the Hudson's Bay Company's trading posts, from which Indians were sent out with abundance of ammunition, and instructions, should they find any white men, to bring them to their forts. The accounts were that at least forty men (the Eskimos find much difficulty in counting any number above five, and even that puzzles them sometimes) were seen dragging a boat or boats on sledges southward, along the west shore of King William Land, and that they had then turned eastward towards the mouth of a large river, which by description could be no other than Back's Great Fish River.

"Later in the spring, when the natives were going to this river to fish, on the first breaking up of the ice, they found what I have described in my report read before the Geographical Society. The whole of this information was sifted over and over again from a number of Eskimos, through my excellent interpreter, whose correctness I was able to prove, by getting through him information from the natives which I found written in the narratives of Ross and Parry. The articles obtained had among them the crests and initials of fifteen of the officers of both ships. For this we were awarded the £10,000 offered by Government. The correctness of my information was five years afterwards wonderfully borne out by that gained by the "Fox" Expedition in 1859, but this information did not extend to the knowledge of any of the crews having reached the mainland ('*noo-nah*'), as in my case.

"The finding of the large quantities of clothing on the north of King William Land, and the boat with two skeletons, guns, etc., by the 'Fox' expedition, on or near its west shore, indicates that the Eskimos had not been there, the reason being, no doubt, that the natives seldom or never travel overland, when they can travel on ice."

Dr. Rae believed, from information obtained at the time, that the Eskimos did not find any of the Franklin ships. On being asked about ships, the natives always reverted to Ross's steamer, the "Victory," abandoned in Boothia Gulf in 1832, all about which he had heard in the course of his expedition in 1846–47. From this vessel the natives had clearly obtained the wood, of which they had enough for all necessary purposes at that time. "My chief reason," writes Rae, "for believing that none of the ships had been found was the fact that, in 1854, the Eskimos were so destitute of wood, that although they had plenty of sealskins to make their small hunting canoes, they had no wood for frames. Now, as 1846 was fourteen years after Ross's vessel was abandoned, and as 1854 was only four years by Eskimo account—actually six years—after the Franklin ships were abandoned, the probability is that had these ships, or even one of them, been found, the natives would have had at least as much wood in 1854 as they had in 1847. The testimony of the 'Fox' expedition of 1854 tends to support this idea, as no large wooden sledges were found, and no wood of a size larger than might have been got from the keel of a boat was seen. . . . I questioned the Repulse Bay Eskimos over and over again about whether any of the ships of the starved white men had been found, but they could tell me nothing, and always went back to the story of the 'Victory,' stating that it was the only vessel from which wood had been obtained. I still believe that this was the ship to which the Eskimos referred when speaking to McClintock in 1859, and that they concealed the locality of the wreck lest he should wish to go there. . . . I may add that the white men, when seen alive by the Eskimos, made the latter understand by signs and a word or two of Eskimo, that they were going to the mainland (*noo-nah*) to shoot deer (*took-took*). All the party except one man, whom the natives took to be a 'chief,' and who had a telescope strapped on his shoulder, were hauling the sledges and boat or boats, and they all looked very thin. The Eskimos also remarked that it was curious that sledges were seen with *the party* when travelling, but none were seen where the dead were, although the boat or boats remained. I pointed out to them that the white men having got close to the mouth of Great Fish River, would require their boat to go up it, but as they did not require the sledges any more, they might have burnt them for fuel. A look of intelligence immediately lit up their faces, and they said that may have been so, for there had been fires. . . . They said also that feathers of geese had been seen, so they had probably shot some of these birds—an evidence that some of the party must have lived until about the beginning of June, the date at which the geese arrive so far north. I may again say, that the Eskimos gave me clearly to understand that the greater part of the dead men were found on the main shore (*noo-nah*), only four or five being found on an island (*kai-ik-tak*). . . . What struck me at the time, as it does still, was the great mistake made by Franklin's party in attempting to save themselves by retreating to the Hudson's Bay territories. We should have thought that the fearful sufferings undergone by Franklin and his companions, Richardson and Back, on a former short journey through these barren grounds, would have deterred inexperienced men from attempting such a thing, when the well-known route to Fury Beach—certainly much more accessible than any of the Hudson's Bay Company's settlements, and by which the Rosses escaped in 1832–33—was open to them. The distance from their ships to Fury Beach was very little greater than that from where Ross's vessel was abandoned to the same place, and Franklin and his officers must have known that an immense stock of provisions still remained at the place where the 'Fury' was wrecked, and where, even so late as 1859, an immense stock of preserved vegetables, soups, tobacco, sugar, flour, etc., still remained (a much larger supply than could be found at many of the Hudson's Bay trading posts); besides, the people would have been in the direct road of searching parties or whalers. The distance to Fury Beach from where the ships were abandoned, roughly measured, is, as nearly as possible, the same as that between the ships and the true mouth of the Great Fish River, or about 210 geographical miles in a straight line. Had the retreat upon Fury Beach been resolved upon, the necessity for hauling heavy boats would have been avoided, for during the previous season (that of 1847), a small sledge party might have been despatched thither to ascertain whether the provisions and boats at the depot were safe and available. The successful performance of such a journey should not have been difficult for an expedition consisting of 130 men who, in the record found in 1859 by McClintock, were reported all well in the spring of 1847."

We have seen that Dr. Rae met his intelligent Eskimo "with the gold cap-band round his head," and learned from him the first trustworthy intelligence respecting the fate of Franklin, on the 21st April 1854, while conducting his party across Boothia to the Castor and Pollux River of Dease and Simpson. The principal object of this journey, it may be necessary to remind our readers, was to complete the discovery and survey of the north coasts of America by exploring the shores between Dease and Simpson's farthest on the south (Castor and Pollux River), with Kennedy's farthest on the north (Bellot Strait). The extraordinary intelligence which Rae had just received respecting the fate of at least one-third of the officers and crews of the "Erebus" and "Terror" had no influence in making the explorer abandon the object of his journey. He still pushed west across Boothia, and at night built his snow-house in lat. 68° 29' N., long. 90° 53' W. The snow-house was built on the frozen bed of a stream which falls into Pelly Bay from the west, in lat. 68° 47', and which Rae afterwards named Becher River. On the following day (the 22d) the travellers marched west for seven or eight miles to Ellice Mountain, then north-east to the east extremity of Simpson Lake, where the camp was pitched. "Our Eskimo auxiliaries," says Rae, "were now anxious to return, being, or professing to be, in dread that the wolves or wolverines should find their *cache* of meat, and destroy it." The explorer therefore paid them liberally, and bade them a friendly farewell. The natives had advised him to follow the chain of lakes that ran in a north-westerly direction and then turned sharply to the southward, and thereafter to follow the stream that flowed westward from the lakes. He learned, however, that to follow this route would lead him too far south; he therefore struck across the land westward, and found himself among a series of hills and valleys in which traces of deer and musk-oxen were of frequent occurrence. At two A.M. on the 26th, after a most laborious walk of eighteen miles across difficult country, he built his snow-hut in lat. 68° 25', long. 93° 4'. On the evening of the same day, Rae, leaving two men to follow at their leisure, set out with the remaining two men to reach the sea at the mouth of Castor and Pollux River. At eight on the morning of the 27th Rae reached the *sea-ice*, in lat. 68° 32' N., long. 93° 44' 48" W., being 3' 38" N., and about 13' E. of Simpson's position of the mouth of the Castor and

Pollux River. "The weather," continues Rae, "was overcast with snow when we resumed our journey at 8.30 P.M. On the 27th we directed our course directly for the shore, which we reached after a sharp walk of an hour and a half. . . . After passing several heaps of stones, which had evidently formed Eskimo *caches*, I came to a collection larger than any I had yet seen, and clearly not intended for the protection of property of any kind. The stones, generally speaking, were small, and had been built in the form of a pillar, but the top had fallen down, as the Eskimos had previously given me to understand was the case. Calling my men to land, I sent one to trace what looked like the bed of a small river, immediately west of us, whilst I and the other man cleared away the pile of stones, in search of a document. Although the cairn contained no document, there could be no doubt in my own mind, or in that of my companion, that its construction was not that of the natives. My belief that we had arrived at the Castor and Pollux River was confirmed when the person who had been sent to trace the apparent stream-bed returned with the information that it was clearly a river. My latitude of the Castor and Pollux River is 68° 28' 37" N., agreeing within a quarter of a mile with that of Simpson,"—which was 68° 28'.

Having reached Simpson's farthest, and even seen the pillar, or, as that explorer names it, the "monument," constructed "in commemoration" of his discoveries on this coast, Rae now prepared to carry out the main object of his expedition by travelling direct north along the Boothian shores to Bellot Strait, and thus connecting the discoveries of Simpson and Kennedy. After a fatiguing march of fifteen hours, during which a distance of thirty miles was traversed, he arrived at the snow-hut of the men that had been left behind. Thence a fresh start was made. An ample stock of provisions and fuel was placed on the two best sledges, and on a third sledge Rae himself dragged his instruments, books, bedding, etc. Among the chief of the Doctor's discoveries on this coast are Murchison River, Shepherd Bay, Bence Jones' Island, Cape Colville, Stanley Island, and Point de la Guiche. Westward from Stanley Island land was discovered at the distance of seven or eight miles, and was named Matheson Island. A more recent discoverer, however, finding that this bold land was really the eastern extremity of King William Island, changed the name to Matheson Mount.

On the 6th May the snow-hut was pitched on Point de la Guiche, in lat. 68° 57' 52", long. 94° 32' 58". One of the men, Mistegan, the Indian hunter, was sent forward six miles north along the coast, where, ascending an elevation, he could see five miles still farther. "The land," says Rae, "was still trending northward, whilst to the north-west, at a considerable distance—perhaps twelve or fourteen miles—there was an appearance of land, the channel between which and the point where he stood was full of rough ice. This land, if it was such, is probably part of Matty Island or King William Land, *which latter is also clearly an island.*" At this point Dr. Rae, having been detained for a number of days by foggy and snowy weather, found the time at his disposal so limited that he could not complete the whole of the survey to Bellot Strait or Brentford Bay without great risk to his party, one of whom had been for many days badly frost-bitten, and had been left behind with a companion. The explorer therefore resolved to retrace his steps without further delay, and having taken possession of his discoveries in the usual manner, he set out on his return journey on the 6th May. On the 11th he reached the spot at which two of his men had been left, and on the same night started for Repulse Bay. Pelly Bay was reached at one A.M. on the 17th, and a snow-house built near the encampment of the 20th April. Traces of Eskimos were observed here, and after supper two men were sent out to follow them up. After eight hours' absence the men returned with ten or twelve native men, women, and children. "From these people," says Rae. "I bought a silver spoon and fork. The initials 'F.R.M.C.' not engraved, but scratched with a sharp instrument on the spoon, puzzled me much, as I knew not at the time the Christian names of the officers of Sir John Franklin's expedition." Committee Bay was reached on the 21st, and Repulse Bay on the 26th May 1854. Dr. Rae found the three men whom he had left in charge here, living in abundance, and on the most friendly terms with the Eskimos, who had pitched their tents near them. "The natives had behaved in the most exemplary manner," writes Rae, "and many of them who were short of food had been supplied with venison from our stores, in compliance with my orders to that effect. It was from this time until August that I had opportunities of questioning the Eskimos regarding the information which I had already obtained, of the party of whites who had perished of starvation, and of eliciting the particulars connected with that sad event, the substance of which I have already stated."

Dr. Rae had still half the original stock of pemmican on hand, together with a sufficiency of ammunition to provide supplies for another winter. The party besides was in excellent health, and he could have procured as many dogs for sledge travelling as would have been required in the event of his deciding to resume the survey of the Boothian coasts in the following year. There was little doubt that a second attempt, therefore, would be successful; "but," says Rae, "I now thought that I had a higher duty to attend to—that duty being to communicate, with as little loss of time as possible, the melancholy tidings which I had heard, and thereby save the risk of more valuable lives being jeopardised in a fruitless search in a direction in which there was not the slightest prospect of obtaining any information." He accordingly embarked with his party on the 4th, and arrived safely at York Factory on the 31st August.

CHAPTER THREE

Anderson's Expedition—No Interpreter To Be Had—Relics Found On Montreal Island—Return Of Expedition.

At the time when the surprising intelligence of Rae's discoveries in Boothia reached us at the close of 1854, England had engaged in a great European conflict. Her troops had been sent to Turkey and the Crimea, and her entire naval force was on active service, either in the Black Sea, the Baltic, or in defence of our own shores and those of our colonies. Yet even at this stormy and eventful period, when the minds of men were thoroughly mastered by the peculiarly distressing details of the Crimean War, the intelligence that the fate of one-third part of the Franklin expedition had been conclusively ascertained, not only won the ear of the entire British people, but created a degree of excitement and painful solicitude which compelled the Government to take some step to follow up the inquiry to which an unmistakable clue had been furnished by Dr. Rae. But what was Her Majesty's Government to do? Neither ships, officers, nor men could be spared when the honour and security of England demanded their presence in the north and in the east of Europe. In this difficulty English ministers had recourse again to the Hudson's Bay Company, whom they requested to organise an expedition to examine Back's Great Fish River in 1855, and endeavour to discover whether any of the Franklin party, who were known to be marching for that river with the object of ascending it and reaching some trading station of the Hudson's Bay Company, still survived. A boat expedition for this purpose was accordingly organised by the Company. If Sherard Osborn's statement be strictly correct, the command of this expedition was offered to Dr. Rae, the most capable traveller and explorer in the Company's service, but was by that officer declined. It seems indeed a little strange that Rae, who was the first to find the clue to the fate of Franklin, should not have endeavoured to follow up that clue and completely solve the Franklin mystery *in the autumn of* 1854, instead of withdrawing at once from the field and returning to England. In fairness, however, to Dr, Rae, it is necessary to explain his declinature of the offered command, and this explanation we are enabled to give, once for all, from original and private documents which the distinguished explorer has kindly placed at our disposal.

"On my return to England in 1854," says Dr. Rae, "I was much blamed by people who knew nothing of the matter for not going in the summer or autumn of 1854 to the place indicated by the Eskimos as the locality where many, in all probability, the last, survivors, of the Franklin crews perished

"This is easily explained. *It was after my return to winter quarters, in* 1854, from our very long sledge journey, that I obtained sufficiently clear information from the natives of the position where the dead white men were found. That they were all dead, and had been so for at least four years, was made evident to me, because I offered immense rewards in guns, kettles, knives, saws, files, etc., and everything that Eskimos most value, if they could tell me of even one man, or the possibility of one man, being alive. But there were actual impossibilities in the way of my getting to the place in the summer or autumn of 1854. In the first place, it is impossible to travel overland when the thawing of the snow is going on. Every stream, however small, is a torrent, and if the banks are at all high, each side has a small precipice or wall of snow that there is no getting over. Apart from this difficulty there was the estuary, many miles wide, of the Great Fish River to cross, which could not have been done without a canoe or a boat, and no such means of conveyance was available. The same difficulty existed as to King William Land (an island). But even had there been a boat or canoe available, the autumn journey could not have been made without exposing my whole party (eight in number) to the almost absolute certainty of starvation. For, as I have already said, we had to depend upon our own guns for our food—shooting deer in their autumn southward migration. But if absent on a journey we could not do this. I the previous season with my own rifle had killed nearly half the game obtained, and as my best men would have had to accompany me on the suggested *autumn expedition*, all the good shots would have been with me, many miles from the passes which the deer frequent at the period of the autumn migration. Then the season was already so far advanced that we would not likely have got back to Repulse Bay until after the formation and setting fast of the sea-ice, so that we could not have pushed southward in our boats. Such were my chief reasons for coming home; but there was another. Four ships of Her Majesty's Navy were in the Arctic Sea searching for the lost expedition in every direction but the right one.

"These ships had orders to remain out for years, a depot ship being sent out annually to be ready in the event of disasters to give aid. I felt that information of my discoveries should be conveyed to these ships as soon as possible, so that they might be recalled. I found them home before me—the men at least—not the ships, for *they* were abandoned. They had remained out only two winters instead of three or four as was anticipated."

It is evident, therefore, that Rae, after his return from Repulse Bay in 1854, could not have undertaken any further exploring that season with the slightest hope of success, while his return to England at the close of that year precluded him from accepting charge of a party in the following spring.

PART TWO

THE MAKING OF AN ARCTIC EXPLORER

Narrative of an expedition to the shores of the Arctic Sea, in 1846 and 1847 is John Rae's detailed account of his first Arctic expedition. Originally published in 1850, it is the most widely known piece of writing he produced, and is included here in its entirety.

CONTENTS

❊ ❊ ❊

CHAPTER ONE — 49
Object and plan of the Expedition—Equipment at York Factory—Boats—Crews—Articles useful in an Arctic Voyage—Breaking up of the ice in Hayes and Nelson Rivers—Departure from York Factory—Progress retarded by the ice—First night at sea—Reflections—Rupert's Creek—Unbroken fields of ice—Broad River—Description of the Coast—Double Cape Churchill—Open sea to the north and north west—Arrive at Churchill—White whales—Mode of catching them—Sir George Simpson's instructions—Stock of provisions.

CHAPTER TWO — 61
Depart from Churchill—A gale—Anchor in Knap's Bay—Land on an island—Esquimaux graves—Visited by Esquimaux—A large river running into Knap's Bay—Nevill's Bay—Corbet's Inlet—Rankin's Inlet—Cape Jalabert—Greenland whales seen—Chesterfield Inlet—Walruses—Cape Fullerton—Visited by an Esquimaux—Reefs—Cape Kendall seen—Ice packed against the shore—Take shelter in an excellent harbour—River traced—Seals—Gale—Ice driven off—Direction of the tides reversed—Whale Point—Many whales seen—Again stopped by the pack—Wager River estuary—Ice drifts—Eddy currents—No second opening into Wager River seen—Enter Repulse Bay—Interview with Esquimaux—No intelligence of Sir John Franklin.

CHAPTER THREE — 73
Receive a visit from a female party—Their persons and dress described—Crossing the Isthmus—Drag one of the boats up a stream—Succession of rapids—North Pole Lake—Find a plant fit for fuel—Christie Lake—Flett Portage—Corrigal Lake—Fish—Deer scaring stones—White wolf—Stony Portage—View of the sea—Exploring party sent in advance—Their report—Long Portage—Difficult tracking—Miles Lake—Muddy Lake—Rich pasturage and great variety of flowers on its banks—Marmot burrows—Salt Lake—Visit Esquimaux tents—Discouraging report of the state of the ice—Esquimaux chart—Reach the sea—Ross inlet—Point Hargrave—Cape Lady Pelly—Stopped by the ice—Put ashore—Find a sledge made of ship-timber—Thick fog—Wolves—Walk along the shore—Remains of musk-cattle and rein-deer—Nature of the coast—Danger from the ice—Irregular rise of the tide—Deer on the ice—Fruitless efforts to proceed northward—Cross over to Melville Peninsula—Gale—Again stopped by the ice—Dangerous position of the boat—Return to starting point—Meeting with our Esquimaux friends at Salt Lake—Deer begun to migrate southward—Walk across the isthmus to Repulse Bay.

CHAPTER FOUR — 88
State of things at Repulse Bay—Determine to discontinue the survey till the spring—Reasons—Party sent to bring over the boat—Fix on a site for winter residence—Ptarmigan—Laughing geese—Eider and king ducks—Visits of natives too frequent—Return of the party sent for the boat—Report the bay more closely packed than before—Preparations for wintering—Fort Hope built—Proceed to North Pole and Christie Lakes to look out for fishing stations—Purchase dogs—Wariness of the deer—Flocks of geese pass southward—Blue-winged and snowgeese—Their habits—Snow-storm—Its effects—Return to Fort Hope—Daily routine—Signs of winter—Deer numerous—Quantity of game killed—Provision-store built of snow—Great fall of snow—Effects of the cold—Adventure with a deer—Visited by a party of natives—Their report of the ice westward of Melville Peninsula—An island said to be wooded—Produce of the chace in October—Temperature—Two observatories built of snow—Band of wolves—A party caught in a snow-storm—Esquimaux theory of the heavenly bodies—Temperature of November—Diminished supply of provisions.

CHAPTER FIVE — 101
Winter arrangements completed—Learn to build snow-houses—Christmas-day—Northpole River frozen to the bottom—1st January—Cheerfulness of the men—Furious snow-storm—Observatories blown down—Boat buried under the snow—Ouligbuck caught in the storm—Dog attacked by a wolf—Party of natives take up their residence near Fort Hope—Esquimaux mentioned by Sir John Ross known to them—Boat dug out of the snow—A runaway wife—Deer begin to migrate northward—A wolf-chase—First deer of the season shot—Difficulty of deer-hunting in spring—Dimensions of an Esquimaux canoe—Serious accident to Ouligbuck—A conjuror—Preparations for the journey northward—Temperature—Aurora Borealis.

CHAPTER SIX — 112
Set out for the north—Equipment of the party—Snow-blindness—Musk-ox—Mode of killing it—Reach the coast near Point Hargrave—Ice rough along shore—Pass Cape Lady Pelly—Unfavourable weather—Slow progress—Put on short allowance—River Ki-ting-nu-yak—Pemmican placed en cache—Cape Weynton—Colvile Bay—High hill—Dogs giving way—Work increased—Snow-house-building—Point Beaufort—Point Sieveright—Keith Bay—Cape Barclay—Another cache—Leave the coast and proceed across the land—River A-ma-took—Dogs knocked up—Lake Ballenden—Harrison islands—Party left to procure provisions—Proceed with two of the men—Cape Berens—Relative effects of an eastern and western aspect—Halkett Inlet—Reach Lord Mayor's Bay—Take formal possession of the country—Commence our return to winter quarters—Friendly interview with the natives—Obtain supplies of provisions from them—View of Pelly Bay—Trace the shore to the eastward—Travel by night—Explore the coast of Simpson's Peninsula—Arrive at Fort Hope—Occurrences during the absence of the exploring party—Character of the Esquimaux Ivitchuk.

CHAPTER SEVEN — 138
Preparations for exploring the coast of Melville Peninsula—Outfit—Leave Fort Hope—Pass over numerous lakes—Guide at fault—Dease Peninsula—Arrive at the sea—Fatigue party sent back to Fort Hope—Barrier of ice—Lefroy Bay—Large island named after the Prince of Wales—Detained by stormy weather—Short allowance—Cape Lady Simpson—Selkirk Bay—Snow knee-deep—Capes Finlayson and Sibbald—Deer shot—A cooking scene—Favourite native relish—Again stopped by stormy weather—Cape McLoughlin—Two men left to hunt and fish—Cape Richardson—Chain of islands—Garry Bay—Prince Albert range of hills—Cape Arrowsmith—Coast much indented—Baker Bay—Provisions fail—proceed with one man—Cape Crozier—Parry Bay—Cape Ellice, the farthest point seen—Take possession—Commence our return—No provisions procured by the men left behind—Short commons—Flock of cranes—Snow-blindness—Arrive at Repulse Bay.

CHAPTER EIGHT — 156
Occurrences at Fort Hope during the absence of the exploring party—Remove from winter quarters to tents—Sun seen at midnight—Build an oven and bake bread—Esquimaux method of catching seals—A concert—Lateness of the summer—A native salmon wear—Salmon spear—Boulders on the surface of the ice—Visited by a native from the Ooglit Islands—His report of occurrences at Igloolik—Indolence of the natives—Ice breaking up—Halkett's air-boat—A storm—The ice dispersed—Prepare for sea.

CHAPTER NINE — 165
Voyage from Repulse Bay to York Factory.

ENDNOTES — 179

APPENDIX — 181
List of Mammalia
———Birds
———Fishes
———Plants
Specimens of Rocks
Dip of the needle and force of magnetic attraction at various stations along the west shore of Hudson's Bay, and at Fort Hope, Repulse Bay
Abstract of Meteorological Journal from September, 1846, to August, 1847.
Figures and Letters used for denoting the state of the weather, &c.

TO
SIR GEORGE SIMPSON,
Governor-in-Chief of Rupert's Land,
THE ZEALOUS PROMOTER OF ARCTIC DISCOVERY,
THIS VOLUME IS INSCRIBED
AS A TRIBUTE OF RESPECT AND REGARD
BY THE AUTHOR.

CHAPTER ONE

Object and plan of the Expedition—Equipment at York Factory—Boats—Crews—Articles useful in an Arctic Voyage—Breaking up of the ice in Hayes and Nelson Rivers—Departure from York Factory—Progress retarded by the ice—First night at sea—Reflections—Rupert's Creek—Unbroken fields of ice—Broad River—Description of the coast—Double Cape Churchill—Open sea to the north and north west—Arrive at Churchill—White whales—Mode of catching them—Sir George Simpson's instructions—Stock of provisions.

It is already well known to those who take an interest in Arctic discovery, that the Hudson's Bay Company intended fitting out an expedition in 1840, which was to have proceeded to the northern shores of America by Back's Great Fish River, for the purpose of tracing the coast between the river Castor and Pollux of Dease and Simpson, and the Strait of the Fury and Hecla, as it was then very generally supposed that Boothia was an island.

The party was to have been commanded by that able and enterprising traveller, Mr. Thomas Simpson, whose indefatigable exertions, in conjunction with those of Mr. Dease, had during the three preceding years effected so much, but his untimely and melancholy fate prevented that intention from being carried into effect, and the survey of the Arctic coast was discontinued for a few years.

When it was determined that the survey should be resumed, Sir George Simpson, Governor-in-Chief of the Company's territories, informed me that a boat expedition to the Arctic Sea was again contemplated, at the same time doing me the honour of proposing that I should take command of it,—a charge which I most joyfully accepted.

The plan of the expedition was different from any that had hitherto been adopted, and was entirely of Sir George Simpson's forming. Its leading features were as follows:—A party of thirteen persons, including two Esquimaux interpreters, was to leave Churchill in two boats at the disruption of the ice, and coast along the western shore of Hudson's Bay to the northward as far as Repulse Bay, or, if thought necessary, to the Strait of the Fury and Hecla. From this latter point the shore of the Arctic Sea was to be traced to Dease and Simpson's farthest discoveries eastward; or, if Boothia Felix should be found to form part of the American continent, up to some place surveyed by Captain or Commander (now Sir John and Sir James C.) Ross.

I started from the Sault de Ste Marie in the latter part of July, 1845, in a canoe which I took on with me as far as Red River, where this frail vessel was changed for a boat, which is better adapted for traversing large sheets of water. We had rather a stormy passage to Norway House, at which place five men were engaged for the expedition, and having brought two with me from the southern department, I required only three more, who I knew could easily be procured at York Factory.

At first there was some difficulty in getting volunteers, as a report had got abroad (set on foot, I believe, by either McKay or Sinclair, guides and steersmen with the expeditions under Sir G. Back and Dease and Simpson), that the whole party, if not starved for want of food, would run the risk of being frozen to death for want of fuel.

After leaving Norway House our progress was slow, the water being very shallow, and our boat rather a heavy drag, for a single crew, over the portages. Two Indians who were engaged, the one to go as far as Oxford House, and the other all the way to York Factory, stipulated that they should do no work on Sunday, to which I readily agreed, thinking that they acted conscientiously, and this I really believe to have been the case with one; but I had some doubts about the sincerity of the other, when I learned that, before leaving us, he had stolen a shirt and blanket from one of the boat's crew.

We arrived at York Factory on the 8th October, during a strong gale of north-east wind with heavy rain and sleet, which had thoroughly drenched us all; in addition to which the men were so bedaubed with mud whilst dragging the boats along shore, that scarcely a feature of their faces could be distinguished.

On landing I was most kindly welcomed by Chief-Factor Hargrave and the other gentlemen of the Factory.

There was little probability of our being able to get to Churchill by water this autumn, nevertheless the boats that had been built for the expedition were launched and put in order for sea. They were fine looking and strong clinker-built craft, 22 feet long by 7 feet 6 inches broad, each capable of carrying between fifty and sixty pieces of goods of 90 lbs. per piece. They were each rigged with two lug sails, to which a jib was afterwards added; under which, with a strong breeze of wind, they were found to work admirably. They were named the "North Pole" and the "Magnet."

We had a continuance of northerly winds until the ice began to form on the river, when it would have been highly imprudent to attempt going along the coast, and I did not wish to run the risk of having our boats stranded, which would have been a very likely occurrence had we put to sea. There was, therefore, nothing to be done but to haul our boats up again; nor did this cause me much disappointment, as I felt pretty certain that, in the following spring, we could advance as fast to the northward as the season of the breaking up of the ice did; and this supposition I afterwards found to be correct.

My attention was now turned to the proper equipment of my party, in which I was most ably assisted by Chief-Factor Hargrave and my friend, Mr. W. Mactavish, who was in charge of York during the temporary absence of the former gentleman, so that, with keeping a meteorological journal—in which the temperature of the air, height of the barometer, force and direction of the winds, and state of the weather were registered eight times a day—and taking observations for latitude, longitude, variation of the compass, and dip of the needle, &c., I had occupation enough on my hands.

Among other articles which I thought might be useful, were a small sheet-iron stove for each boat, a set of sheet-iron lamps for burning oil after the Esquimaux fashion, some small kettles (commonly called conjurors) having a small basin and perforated tin stand for burning alcohol, a seine net, and four small windows, each of two double panes of glass. An oiled canvass canoe was made, and we also had one of Halkett's air boats, large enough to carry three persons. This last useful and light little vessel ought to form part of the equipment of every expedition.

On the 30th April, 1846, that harbinger of spring, the Canada goose, was seen, and so early as the 5th May the ice in Hayes River commenced breaking up, but it was more than a month after this date before the Nelson or North River opened. At length, on the 12th June, it was reported that a passage was practicable, and everything was got in readiness for making a start on the following day.

The crews of the boats were divided as follows:—

NORTH POLE
John Rae.
John Corrigal, Orkneyman, Steersman.
Richard Turner, half-breed, Middleman.
Edward Hutchison, Orkneyman, ditto.
Hilard Mineau, Canadian, ditto.
Nibitabo, Cree Indian, ditto and hunter.

MAGNET
George Flett, Orkneyman, Steersman.
John Folster, ditto, Middleman.
William Adamson, Zetlander, ditto.
Jacques St. Germain, Canadian, ditto.
Peter Matheson, Highlander, ditto.

All these men had the same wages, namely, £40 per annum, with the promise of a gratuity in the event of good conduct.

The lading of each of the boats, including the men's luggage, amounted to about seventy pieces; and with this cargo they were quite deep enough in the water and very much lumbered—so much so that, to allow room for pulling, a quantity of the cargo had to be displaced.

On the 13th June, after bidding farewell to our kind friends at York, and receiving a salute of seven guns and three hearty cheers, we set sail with a light air of fair wind. We had not proceeded more than a mile down the river, when the wind chopped round directly in our teeth, and blew a gale. As I could not think of turning back, we were speedily under close-reefed sails, turning to windward; the wind and tide were going in opposite directions, and there was an ugly cross sea running, which caused us to ship much water over both the lee and weather side. After a couple hours of this work we gained sufficient offing to clear the shallows, which lie for some miles out from the point of Marsh, (this being the name of the N.E. extremity of York Island), and stood across towards the north shore of the Nelson River. The men in the Magnet, having erroneously carried on too great a press of canvass, were left a mile or two astern. As we advanced the wind gradually abated, and we soon fell in with quantities of ice driving along with the current, through which we had much difficulty in finding a passage.

We made the land near Sam's Creek, and it being now calm, and flood tide strong against us, we cast anchor close to the shore between 9 and 10 o'clock. The night was beautiful, and, as all my men had gone to sleep, nothing interrupted the stillness around but the occasional blowing of a white whale, the rather musical note of the "caca wee" (long-tailed duck), or the harsh scream of the great northern diver. Yet I could not close my eyes. Nor was this wakefulness caused by the want of comfort in my bed, which I must own was none of the most inviting, as it consisted of a number of hard-packed bags of flour, over which a blanket was spread, so that I had to accommodate myself in the best way I could to the inequalities of the surface. To a man who had slept soundly in all sorts of places—on the top of a round log, in the middle of a swamp, as well as on the wet shingle beach, such a bed was no hardship; but thoughts now pressed upon me which during the bustle and occupation of preparation had no time to intrude. I could not conceal from myself that many of my brother officers, men of great experience in the Indian country, were of opinion that we ran much risk of starving; little was known of the resources of that part of the country to which we were bound, and all agreed that there was little chance of procuring fuel, unless some oil could be obtained from the natives. Yet the novelty of our route, and of our intended mode of operations, had a strong charm for me, and gave me an excitement which I could not otherwise have felt.

14th.—As there were great quantities of ice along the shore to the northward of us, I let the boats take the ground, so that this morning they were high and dry on the mud, the water being a mile or two outside of us, and we as far from the high-water mark.

As the Goose Hunt House (a small hut where one of the Company's servants and some Indians go every spring and autumn to shoot and salt geese,) was at no great distance, I visited it, but found that the people had taken their departure for the Factory—a certain sign that the geese and ducks had gone farther north. Numbers of the Hudsonian godwit (*limosa Hudsonica*) were flying about, apparently intending to breed in the neighbourhood.

The boats floated at a quarter after 10 A.M., and we got under weigh with a fine light breeze from the S.E. The temperature of the air was 62° and the water 40°. There were many pieces of ice floating about, and a great quantity close-packed about half a mile outside. At mid-day we were in latitude 57° 25' 93" N. After running by Massey's patent log for 10¾ miles north, we were stopped by ice at a few minutes after 1 P.M., when we made fast to a large grounded mass, which protected us from the smaller floating pieces as long as the tide was ebbing; but as soon as the flood made, it required all our exertions to prevent the boats being damaged. We now found the great advantage of some sheet copper that had been nailed on their bows, as it completely protected them from being chafed. At 11 next forenoon, finding our situation rather dangerous, as soon as the tide flowed far enough, we pushed inshore, and beached the boats on a fine smooth surface of mud and gravel. With the exception of a heavy shower of rain at 6 A.M., the weather continued fine all day, but the sky was too cloudy to permit any observations to be made.

On the 16th we advanced only 1½ miles. The temperature of the air 42° and the water 34°. By an azimuth of the sun the variation of the compass, 10° 54' east, was obtained.

As it was only at, or near, high water that we could make any progress, we crept along shore about four miles during the morning's tide, and in the evening we put into Rupert's Creek, which afforded us good shelter, and also fresh water, of which we were getting rather short. A fresh breeze from the east brought in much ice, which completely blockaded our harbour. The morning of the 18th was very fine, but the easterly wind still continued, and such was the effect produced by it that not a spot of open water was to be seen. The latitude 57° 32' 18" was observed, and an observation of the sun's azimuth yesterday gave the variation of the compass 9° 56' E. Some partridges (*tetrao saliceti*), ducks, and a flat-billed phalarope (*P. fulicarius*) were shot.

19th.—The ice having become somewhat more open during the night, we left the creek at 4 A.M., and ran 32½ miles before a fine breeze of S.E. wind, through lanes of open water, as nearly as possible in a N.N.E. course. Large unbroken fields, on which numbers of seals were lying, now opposed our further progress. At high water next morning, we set forward among ice so closely packed, that we were obliged to open a passage by pushing aside the smaller pieces; we thus gained between two and three miles and reached Broad River. We lay here during the remainder of the day, which was too cloudy for a meridian observation, but in the evening an amplitude of the sun gave variation 12° 19' east. The dip of the needle was 81° 46' 4".

The morning's tide of the 21st advanced us nearly three miles. Our new position was found to be in latitude 58° 9' 51" N.; the latitude of Broad River must therefore be 58° 7' N. A strong breeze of S.S.W. wind had driven out some of the ice, so that, with the aid of sails and poles, we gained 12 miles more northing in the evening.

From the 22d to the 24th we continued to creep along-shore, but our progress was very slow, 19 miles being, at the highest estimate, as much as we gained. We were, however, killing ducks of various kinds, and collecting eggs enough to keep us in food. A deer was also shot by Nibitabo on the 22d, and on the 24th I procured from a high mound of ice, where it was feeding, what appeared to be a Canada mithatch (*sitta Canadensis*). The skin was preserved, and is with other specimens in the Honourable Hudson's Bay Company's warehouse in London.

On the 25th we lay all day in a small creek, which afforded us a safe harbour.

The wind, which had yesterday blown a strong gale from the N.E., shifted round to W., which gave us some hopes of an opening to seaward. In the evening much ice drove out with the ebb. The latitude of our position by reduction to the meridian was 58° 31' N.

26th.—This morning we were fortunate enough, after a great deal of trouble, to get the boats into comparatively open water, and as the wind was moderate from E.S.E. we threaded our way, through narrow channels and openings, until opposite Cape Churchill. At 3 P.M. we doubled the cape, and to our great joy found an open sea to the north and north west of it.

The whole of the coast between Nelson River and Cape Churchill is low and flat, with not a single rock in situ. There are, however, a number of boulder stones of granite, and debris of limestone, to be seen.

There are numerous lakelets near the shore, the banks of which form the favourite breeding places of the Canada goose, the mallard, pintail, teal, scaup, and long-tailed ducks, great northern diver,[1] and the Arctic tern. The phalaropus hyperboreus is also very numerous—so much so that I could have shot twenty in half-an-hour. The female of this phalarope and of the P. fulicarius is considerably larger, and has much finer markings on its plumage, than the male, the colours being much brighter.

As we sailed along shore to the westward, the land gradually became more high and rocky, and there were many ridges of stones lying off several miles from the beach, among which we had some trouble in threading our way, the navigation being rendered still more difficult by a thick fog.

We arrived at the mouth of Churchill River at 3 A.M. on the 27th, but as the tide was ebbing we could not stem the current, so that we did not reach the Company's Fort, situated on the west bank of the river and about five miles up, until half-past six, when I was most kindly welcomed by my friend Mr. Sinclair, chief trader, the gentleman in charge, who had not expected to see us so early.

My letter of instructions had not yet arrived, so that we took advantage of the delay thus occasioned to have the boats unloaded, some slight repairs effected, and the cargoes examined and dried. I determined on leaving here some tobacco, salt, and one or two other articles that were not absolutely essential, supplying their place with pemmican and flour. Some observations for the dip of the needle gave mean dip 84° 47' 3". The variation of the compass 12° 29' east, and the latitude of the Establishment 58° 44' 12" were found, and the mean time of 70 vertical vibrations of the needle in the magnetic meridian was 148".

The people of the fort were busy killing white whales, great numbers of which come up the river with the flood tide. The mode of taking them is very simple. A boat, having a harpooner both at bow and stern, sails out among the shoal, and being painted white, it does not alarm them; they approach quite close, and are thus easily struck. When harpooned they do not run any great distance in one direction, but dart about much in the way that a trout does when hooked.

On the evening of the 4th July the anxiously expected instructions arrived from Red River, via York Factory. The following is a copy of them:—

Red River Settlement, 15th June, 1846.

Sir,

You are aware that the grand object of the expedition which has been placed under your direction is to complete the geography of the northern shore of America, by surveying the only section of the same that has not yet been traced, namely, the deep bay, as it is supposed to be, stretching from the western extremity of the Straits of the Fury and Hecla to the eastern limit of the discoveries of Messrs. Dease and Simpson.

2. For this purpose you will proceed from Churchill with the two boats, and twelve men that have been selected for this arduous and important service, losing not a moment, at least on your outward voyage, in examining such part of the coast as has already been visited and explored. In a word, you will reach, with as little loss of time as possible, the interesting scene of your exclusive labours.

3. In prosecuting the survey in question, you will, as a matter of course, endeavour to ascertain as accurately as circumstances may permit, without occasioning any serious delay, the latitudes and longitudes of all the most remarkable points within the range of your operations, and also the general bearing and extent of all the intermediate portions of coast, embodying the whole at the same time in the form of a chart, or rather of the draft of a chart, from day to day.

4. But in addition to this, your principal and essential task, you will devote as much of your attention as possible to various subordinate and incidental duties. You will do your utmost, consistently with the success of your main object, to attend to botany and geology; to zoology in all its departments; to the temperature both of the air and of the water; to the conditions of the atmosphere and the state of the ice; to winds and currents; to the soundings as well with respect to bottom as with respect to depth; to the magnetic dip and the variation of the compass; to the aurora borealis and the refraction of light. You will also, to the best of your opportunities, observe the ethnographical peculiarities of the Esquimaux of the country, and in the event of your wintering within the Arctic Circle, you will be careful to notice any characteristic features or influences of the long night of the high latitudes in question. These particulars, and

such others as may suggest themselves to you on the spot, you will record fully and precisely in a journal, to be kept, as far as practicable, from day to day, collecting at the same time any new, curious, or interesting specimens, in illustration of any of the foregoing heads.

5. In order to provide against the probable necessity of requiring two seasons for your operations, you will take with you all the provisions that your boats can carry, with such shooting, hunting, and fishing tackle as may enable you to husband your supplies. I need hardly mention medicines and warm clothing among the necessaries of your voyage, for, in full reliance on your professional zeal and ability, I place the health of your people, under Providence, entirely in your hands.

6. In the event of wintering in the country, you will cultivate the most friendly relations with the natives, taking care, however, to guard against surprise. For this purpose you will repeatedly and constantly inculcate on your men, collectively and individually, the absolute necessity of mildness and firmness, of frankness and circumspection.

7. If, in the event of your being unable to accomplish the whole of your task in one season, you see ground for doubting whether the resources of the country are competent to maintain the whole of your people, you will in that case send back a part of them to Churchill with one of the boats. For the remaining part of your men you cannot fail to find subsistence, animated as you and they are by a determination to fulfil your mission at the cost of danger, fatigue, and privation. Wherever the natives can live, I can have no fears with respect to you, more particularly as you will have the advantage of the Esquimaux, not merely in your actual supplies, but also in the means of recruiting and renewing them.

8. During the winter you will pursue the various objects of the expedition by making excursions, with a due regard, of course, to safety, on the snow or on the ice; and at the close of your second season, after having accomplished the whole of

your task, you will return according to your own discretion, either by your original course or by Back's Great Fish River, keeping constantly in view, till you reach Churchill or Great Slave Lake, the general spirit of these your instructions.

9. In conclusion, let me assure you that we look confidently to you for the solution of what may be deemed the final problem in the geography of the northern hemisphere. The eyes of all who take an interest in the subject are fixed on the Hudson's Bay Company; from us the world expects the final settlement of the question that has occupied the attention of our country for two hundred years; and your safe and triumphant return, which may God in His mercy grant, will, I trust, speedily compensate the Hudson's Bay Company for its repeated sacrifices and its protracted anxieties.

I remain,
Sir, &c.
(Signed) G. Simpson.
John Rae, Esq.
Churchill, Hudson's Bay.

The boats were loaded during the night, and at 6 A.M. were sent down to the old stone fort at the mouth of the river, where they were to wait for me a few hours. Besides an abundant supply of ammunition, guns, nets, twines, &c. for our own use, and various articles for presents and to barter with the Esquimaux, we had on board

20 bags pemmican, about 90 lbs. each,
2 ditto grease,   "   90 lbs.   ".
25 ditto flour, each 1 cwt
4 gallons of alcohol for fuel,

with a good stock of tea, sugar, and chocolate, but only four gallons of brandy and two gallons of port wine, as I was well aware of the bad effects of spirits in a cold climate. Considering that we were to be absent fifteen or perhaps twenty-seven months, our quantity of provisions (amounting in all to little more than four months' consumption at full allowance) was not very great.

CHAPTER TWO

Depart from Churchill—A gale—Anchor in Knap's Bay—Land on an island—Esquimaux graves—Visited by Esquimaux—A large river running into Knap's Bay—Nevill's Bay—Corbet's Inlet—Rankin's Inlet—Cape Jalabert—Greenland whales seen—Chesterfield Inlet—Walruses—Cape Fullerton—Visited by an Esquimaux—Reefs—Cape Kendall seen—Ice packed against the shore—Take shelter in an excellent harbour—River traced—Seals—Gale—Ice driven off—Direction of the tides reversed—Whale Point—Many whales seen—Again stopped by the pack—Wager River estuary—Ice drifts—Eddy currents—No second opening into Wager River seen—Enter Repulse Bay—Interview with Esquimaux—No intelligence of Sir John Franklin.

Having taken on board Ouligbuck and one of his sons as Esquimaux interpreters, and bid adieu to Mr. Sinclair, who, during our stay, had omitted nothing that could in any way tend to the comfort of the party, we set sail at 11 o'clock on the 5th July with a light air of N.N.E. wind, and stood to the westward across Button's Bay. The weather was fine, and to enliven the scene numbers of white whales were seen sporting about, and sometimes coming within a few yards of the boats. The men were all in excellent health and spirits, there not being a melancholy look nor a desponding word to be seen or heard among them.

At 3.30 P.M. we passed Pauk-a-thau-kis-cow River, and the wind having freshened and shifted round to the S.E. we had run upwards of forty miles before 10 o'clock. The temperature of the air was 49°, and of the water 50°, thus showing that there was little or no ice in the neighbourhood.

The night being fine we continued under sail, the crews being divided into two watches. The land had now become much lower than it was about Churchill, and the coast very flat; so that it was necessary to keep six or eight miles from the land when the tide was out; and even then, although the boats drew only two and a half feet water, there was little enough for them. The bottom was of mud, sand, or shingle, with every here and there a large boulder stone, some of them ten or twelve feet high.

Early on the morning of the 6th three Esquimaux came off in their kayaks, and although we were going at the rate of four miles an hour they easily overtook us. As they were going towards Churchill, I sent a few lines to Mr. Sinclair by them.

Our latitude at noon was 60° 17' 59" N. Thermometer in air 49°, in water 45°. The total distance run, measured by Massey's log, was ninety-five miles, which agreed very nearly with our latitude, the difference being easily accounted for by the circumstance that the ebb tide runs much stronger to the northward than the flood does in an opposite direction.

In the afternoon there was a strong breeze, which, although fair, was rather too much onshore and raised a heavy sea. At 5 P.M., having run twenty-five miles since noon, we got into shallow water, and although the heads of the boats were immediately turned to seaward, the ebb tide was too quick for us, and we got aground, being ten miles from the main shore. Five miles N.W. of us there was a small but steep island, on the E. side of which there was still a deep snow drift. By a meridian altitude of the moon our latitude was 60° 47' 24" N.

The following morning we floated at 2 A.M., and with a strong breeze from S.E. stood on our course. The weather looked threatening, and we had not been long out before the wind increased to a gale, and the sea rose in proportion. The boats fully realised the good opinion we had of them, but being so deeply laden the sea broke frequently over them, and kept us continually baling; at last the Magnet shipped a heavy sea, and the steersman, either from losing his presence of mind or from not knowing how to act, allowed the boat to broach to. Fortunately no other sea struck her whilst thus placed, else both she and the crew must inevitably have been lost. I here saw the benefit of the precaution I had taken to have some Orkneymen with me, for it was evident the others (although as good fellows as could possibly be wished) knew nothing about the management of a boat in such weather.

I was loath to lose so fine an opportunity of getting on, but it would have been recklessness to attempt proceeding. We accordingly ran in towards Knap's Bay, which was nearly abreast of us, and were soon anchored in a snug cove under the lee of the largest island in the bay. It was well that we put in here, for the wind in a short time increased to a perfect storm with heavy rain.

On a neighbouring island some miles to the south of us, many Esquimaux tents were seen, but we could not discover if they were inhabited.

Notwithstanding the rain I took my gun and made a tour of the island. It is about two miles long, a quarter of a mile broad, and not exceeding 100 feet in height, being covered with a scanty vegetation, and thickly strewn in many places with fragments of granite.

I met with a great many Esquimaux graves, the bodies being protected from wild animals by an arch of stone built over them. We found a number of spear-heads, knives, &c. placed in some of these heaps of stones, but the Esquimaux do not, I believe, destroy all the property of the deceased, as is common among most tribes of Indians.

Tracks were seen of a large white bear which had evidently been feasting on the eggs of various wildfowl that breed here; among these I noticed the eider duck (*fuligula mollissima*), the long-tailed duck (*fuligula glacialis*), and the black guillemot (*uria grylle*).

In the evening, when the wind had somewhat moderated, we were visited by five Esquimaux from the tents before mentioned; they each received a piece of tobacco, of which they are remarkably fond; and one of them promised to carry or forward to Churchill a letter which I addressed to Sir George Simpson. In a net that we had set, a salmon weighing 10 lbs. was caught. A large and deep river empties its waters into this bay; its course is about due east, and it abounds with salmon, seals, and white whales, being consequently a favourite resort of the natives. The rise of the tide was thirteen feet. When about to go to bed I found my blankets quite wet by the seas that washed over me in the morning; this, however, made them keep out the wind better, and did not certainly affect my rest.

The following day was more moderate, but it was 2 P.M. before we could venture out of our harbour. By observation the latitude 61° 9' 42" N., and the variation of the compass 7° 48' east were obtained; the dip of the needle being 86° 18' 3" N.

At 4 A.M. on the 9th the wind went round so far to the east that we could not lie our course; it rained heavily, but the wind became more favourable, and we stood over towards the north shore of Nevill's Bay. The temperature of the water at mid-day 37°, air 44°; latitude by observation 61° 55' 40" N.

We passed among many small islands, the resort of great numbers of the birds already mentioned, which we used as food (although not very palatable) to save our pemmican. I also noticed a few of the foolish guillemot (*uria troile*), the first we had met with.[2] At half-past five, it being calm, we landed on a small island to get some water; we found a few Hutchins geese here, one of them having a brood of young with her. These appear to have taken the place of the Canada goose, as I have not seen any of the latter lately. At 8 o'clock, it still being calm, we pulled up towards the north point of Nevill's Bay, which bore east of us. No ice was to be seen, but there were numerous patches of snow on the main shore N.E. of us, distant 10 or 11 miles.

I saw a young shore lark and a young snow bunting, both able to fly. There are quantities of red, grey, and blue granite in this island, variegated with quartz.

The shores had now become steep and rugged, the whole coast being lined with bare primitive rocks.

After breakfast next morning we pulled round the east end of some rocks near which we had lain at anchor during the flood tide, and kept on our course across Whale Cove. Some small pieces of ice were seen floating about, the thermometer in the shade 55°, water 36°. A fog, which had been thick all the morning, cleared up at half-past ten, and we saw some islands at no great distance right a-head, and a smoke a few miles inland on our beam, probably made by Esquimaux, but we could not see any tents. Our latitude by observation was 62° 11' 23" N. Temperature of air 55°, of water 37°.

The weather was very variable, with calms and light breezes alternately. At a little after 7 in the evening we were off the south point of Corbet's Inlet. It rained hard almost all night; we, however, continued our course, and at 7 A.M. got among a number of reefs and islands that lie near the south point of Rankin's Inlet. In attempting to pass between two of these our boat got aground, and as the tide was ebbing she could not be shoved afloat again; but, as the greater part of the cargo was carried on shore before the water fell very far, no damage was done. An excellent observation of the sun gave latitude 62° 35' 47" N., variation 6° 6' W., Marble Island bearing east by compass. The black guillemot was in such numbers here that four or five were killed at one shot. Many eggs were collected, and one nest was found having two eider and three long-tailed ducks' eggs in it. The eider had possession, but whether the birds had a mutual understanding, or whether the stronger had driven out the weaker possessor, it is difficult to say.

At 4 P.M. we floated and ran across the inlet, the traverse being 15 miles. We landed at its north point, as the wind and tide were both against us. There were numerous signs that this place is often visited by the Esquimaux; the bones of various animals and the remains of some stone "caches" being every where visible. A little before midnight a deer was shot by Corrigal. During a walk I fell in with a large white owl (*strix nyctra*). As is usually the case it was very shy, and could not be approached within gun-shot.[3]

The rise of the tide was 14 feet.

At half-past two A.M. on the 13th we landed at Jalabert. The morning was delightful, being quite calm with a sharp frost. While we lay here waiting the change of the tide, Ouligbuck shot a fine large buck. Many seals were sporting about, and a shoal of salmon were seen swimming close to the beach. Having taken on board our venison, we pulled with the tide now in our favour. We saw upwards of a dozen Greenland whales, all apparently busy feeding, some of them very large. At noon we were in latitude 63° 6' 14" N. The variation of the compass 8° 52' W. In the evening we passed Chesterfield Inlet. Great numbers of rocks lie out fully eight miles from the shore on its north side. The wind continued fair and moderate all night, and at 6 in the morning, when in the large bay S.W. of Cape Fullerton, a single Esquimaux visited us in his kayak. He had been at Churchill last year, but did not intend to go thither this season, although he had a number of wolf, fox, and parchment deer skins at his tent. A present of a knife and a piece of tobacco made him quite happy, and he left us shouting so loudly as to show that his lungs were in good order. The party to which he belongs consists of ten families, their hunting-grounds being situated on the borders of Chesterfield Inlet, where they spear a great number of deer whilst swimming across in the autumn. At some distance inland, woods are found. A number of walruses were observed lying on a small ridge of rocks. They were grunting and bellowing—making a noise which I fancy would much resemble a concert of old boars and buffaloes. We did not disturb their music. Obtained a meridian observation of the sun, which gave latitude 64° 3' 42" N. As the refraction was great and the natural horizon used, this is probably erroneous; if it is not, Cape Fullerton is not properly laid down in the charts, being too far to the south. Temperature of the air 58°, water 41°.

When doubling Cape Fullerton, we were obliged, by the numerous granite reefs, to keep six or seven miles from the mainland. At 7 in the evening we landed to replenish our water casks, and had an unsuccessful chase after two deer. The horizon being clear, I saw Cape Kendall on Southampton Island, bearing S.E. by S. magnetic.

15th.—We made but little progress last night, there being no wind. The weather was rather cold, the thermometer standing at 40°, and the water being only 4° above the freezing point indicated the proximity of ice. A short time afterwards a large *pack* was seen about five miles distant. On approaching nearer, we found that it extended along shore as far as the eye could see. At 2 P.M. we ran inshore, and took shelter under some grey-coloured granite rocks twenty feet high. Deer being noticed at no great distance, two or three sportsmen went after them, and succeeded in shooting a doe. A very large whale was observed.

Finding our present position far from being a safe one, at high water we pushed along shore among masses of ice during a thick fog, and entered an inlet which opportunely presented itself, and which proved to be an excellent harbour about 200 yards wide, from four to six fathoms deep, and nearly four miles long. The bottom being sand and mud would afford excellent anchorage for much larger craft than ours. As there were many seals swimming about, I was led to infer that salmon or trout were abundant; two nets were put down, but no fish were caught.

During a two days' detention here I traced, for eight miles, the course of a considerable river which empties its waters into the inlet. I found it to be a succession of rapids and deep pools, and running as nearly as possible in a S.S.E. course. Near its mouth upwards of thirty seals were lying basking in the sun; a ball fired among them sent the whole party walloping into the water at a great rate, more frightened, however, than hurt. One of the men had accompanied me, and during our walk we met with a hen partridge (*tetrao rupestris*) and her brood. I have seen many birds attempt to defend their young, but never witnessed one so devotedly brave as this mother; she ran about us, over and between our feet, striking at our hands when we attempted to take hold of her young, so that she herself was easily made prisoner. Although kept in the hand some time, when let loose again she continued her attacks with unabated courage and perseverance, and was soon left mistress of the field, with her family safe around her.

We were fortunate in finding some willows fully an inch in diameter, which were far superior for fuel to the sea-weed and short heath we had been using for the last two days.

Hutchins geese breed here in numbers, and as no Canada geese were seen, I presume that they do not usually come so far north along the coast. The shores have a very rugged appearance, there being numerous high ridges of primitive rocks running far out into the sea in an east and west direction, the line of stratification dipping to the south at an angle of 75° with the horizon. In many places these rocks were thickly studded with small garnets. The rise and fall of the tide was 13½ feet.

During the whole of the 16th the weather was cloudy, and it rained heavily all night, but on the 17th the wind increased to a gale, the sky cleared up, and a satisfactory observation was obtained by the artificial horizon, which placed us in latitude 64° 6' 45" N. As we were more than ten miles north of the situation where I had observed the latitude on the 14th, the difference between the latitude obtained then and that of our present situation shews the uncertainty of observations made with the natural horizon when there is much refraction, or when there is ice in the neighbourhood. The variation of the compass was 20° 10' W. The gale continued all day, and being from the westward much ice was driven off shore.

18th.—Last night the wind moderated a little, but about 2 A.M. it blew more strongly than before. The forenoon was sufficiently fine to permit me to observe the dip of the needle 86° 36' 5" N.

In the afternoon, when collecting plants, I discovered some willows of a larger growth than those we had before found, and I carried a load of them to the boat. In the evening there was no ice to be seen either along shore or in the offing, but it still blew too hard for us to get under weigh. The temperature of the air to-day varied from 50° to 55°. Just as I had turned in for the night, it was reported that two white bears were close at hand. I immediately got up, and set off "sans culottes" to have a share of the anticipated sport, when I soon discovered that two harmless deer in their winter coats had been mistaken for bears. It was high water to-day at 11h. 40m. A.M., the rise being 15 feet. By this it will appear that 3 o'clock is the time of high water at full and change of the moon.

At 3 next morning, the wind having moderated, we started, and ran along shore at a fine rate for ten miles, but here the coast turning more to the westward we could not lie our course, and were compelled to put ashore until the flood tide made; for it was found that, contrary to what we had previously experienced, the current ran to the northward during the flow of the tide, and in an opposite direction during the ebb, this being probably caused by the strait north of Southampton Island being blocked up with ice. After an hour's stay we got under weigh again at a few minutes after seven, and turned to windward. Our latitude at noon was 64° 20' 51" N. It now fell calm; but this had not continued more than half an hour before a light breeze sprung up from the east, and at 1 P.M. we passed Whale Point. A great many whales were seen to-day, and one of them was swimming amongst a large flock of king ducks, apparently amusing itself with the confusion that it caused when rising to breathe. Temperature of the air 50°—water 38°.

20th.—It being calm for some time during the night, we came to anchor whilst the tide was against us; but at 6 A.M. we again continued our route. There was much ice lying along the shore of Southampton Island, its proximity being indicated by the temperature of the water (35°) this morning. Some more large whales were noticed. The ice was again too close packed to permit us to advance; we therefore landed, and the latitude 64° 56' 33" N., and the variation of the compass 36° 13' W., were observed. The musquitoes were very numerous and troublesome, but, nevertheless, the sportsmen succeeded in shooting five deer.

On the 21st and 22d we had a continued struggle amongst heavy and close-packed ice until we reached Wager River Estuary, where we were detained all day by the immense quantities driving in with the flood and out again with the ebb tide, which ran at the rate of 7 or 8 miles an hour, forcing up the floes into large mounds, and grinding them against the rocks, with a noise resembling thunder.

During the ebb tide the eddy currents once or twice brought in the ice with great force, which would have smashed our boats, as they lay in rather an exposed situation along the face of some steep rocks, had it not fortunately taken the ground before it reached us. During our stay, a meridian observation of the sun by artificial horizon gave latitude 65° 15' 36" N., variation 48° 13' W.

23rd.—There was a thin coat of ice on the water this morning, the temperature of which at midnight was 2° below the freezing point, that of the air 36°. As our position was far from safe, we were kept on the alert all night, and got under weigh at half-past three, for the purpose of finding some safer harbour. To get to a small bay a mile and a half to the west of us, we had more than once to pull for our lives, as the eddy currents already spoken of caused such sudden and uncertain movements among the ice that there was no telling on what side we were to expect it. With much difficulty we entered our harbour, and pulled half a mile up, so as to be safe from the ice, which we had reason to expect would come in with the flood. The latitude of our new anchorage was 65° 16' 8" N. This is the most northerly point on the south side of Wager River, which appears to be not very correctly laid down in the charts. The channel is not more than four or five miles broad. In the evening, being wearied with delay, as soon as the flood tide slacked, we pushed out into the stream, and when in mid-channel had the advantage of a fine breeze, which enabled us to stem the current that still ran at the rate of five miles an hour. The boats had some narrow escapes, and the Magnet received a severe squeeze, but fortunately sustained no injury, and we were soon in safety on the north side of the channel.

24th.—Having pulled along shore all night, we cast anchor at half-past five this morning to take breakfast and give rest to the men. Our course since crossing Wager River had been among a number of small rocky islands, between which we had some difficulty in threading our way, but we did not see any signs of a second opening into Wager Bay, although a sharp look-out was kept. A light air of fair wind springing up, we got under weigh at a few minutes before 8, and stood on to the northward, the ebb tide again running with us. At mid-day the temperature of the air was 45°, water 32°.

In the afternoon the breeze increased, and at a quarter-past seven we rounded Cape Hope, and ran into Repulse Bay. By an amplitude of the sun whilst setting, the variation of the compass 62° 40' W.

was obtained. As soon as we passed the Cape a great change in the temperature of the air and water was observed, the former being 56°, and the latter 46°.

25th.—We continued under sail all night, and at 6 in the morning were within seven miles of the head of the Bay, and cast anchor between a small island and the shore to get some fuel and cook breakfast. Our latitude was 66° 26' 57" N. Variation of compass 59° 10' W.

In the afternoon, the wind being ahead, we plied to windward, and when entering Gibson's Cove, observed with much joy four Esquimaux on the shore. I immediately landed near them, and taking Ouligbuck's son with me as interpreter, joined the party, and calling out Texma (peace), shook hands with them. They were at first in great fear, and appeared half inclined to run away, but on our kind intentions towards them being explained they became quite at ease, chatting and laughing as if we had been old acquaintances. They were good-looking, of low stature, and much more cleanly than those in Hudson's Straits. Their dresses were made of deer skin, of the form so often described, the coat having a long tail somewhat resembling that of an English dress coat. Their legs were encased in waterproof boots made of seal-skin, and they all wore mittens, which they seldom took off their hands. There were two of them middle-aged, Oo-too-ou-ni-ak (who had a formidable beard and whiskers) and Kir-ik-too-oo; the other two were lads from eighteen to twenty years of age; and we were soon after joined by a fine young fellow with ruddy cheeks and sparkling black eyes, having an expression of exceeding good humour in his laughing countenance. Our new friend wore round his head a narrow leather band of deer-skin ornamented with foxes' teeth, and appeared to be somewhat of a dandy in his own estimation. None of the party had ever visited Churchill, and they had neither heard nor seen anything of Sir John Franklin. From a chart drawn by one of the party, I was led to infer that the sea (Akkoolee), to the west of Melville Peninsula, was not much more than forty miles distant in a N.N.W. direction, and that about thirty-five miles of this distance was occupied by deep lakes; so that we would have only five miles of land to haul a boat over—a mode of proceeding which, even had the distance been much greater, I had intended adopting, in preference to going round by the Fury and Hecla Straits.

A small river empties its waters into the Bay within a hundred yards of the place where we landed: this stream, up which the boat was to be dragged, issues from one of the lakes through which we had to pass. Leaving all the men but one to unload the boats, I went some miles inland to trace our intended route. After walking about five miles along the stream already mentioned (the current in which was very strong), we arrived at the first lake, a long and narrow body of water, having steep and in some places rocky banks, which we traced for two miles, and returned late in the evening to our companions.

CHAPTER THREE

Receive a visit from a female party—Their persons and dress described—Crossing the Isthmus—Drag one of the boats up a stream—Succession of rapids—North Pole Lake—Find a plant fit for fuel—Christie Lake—Flett Portage—Corrigal Lake—Fish—Deer-scaring stones—White wolf—Stony Portage—View of the sea—Exploring parties sent in advance—Their report—Long Portage—Difficult tracking—Miles Lake—Muddy Lake—Rich pasturage and great variety of flowers on its banks—Marmot burrows—Salt Lake—Visit Esquimaux tents—Discouraging report of the state of the ice—Esquimaux chart—Reach the sea—Ross inlet—Point Hargrave—Cape Lady Pelly—Stopped by the ice—Put ashore—Find a sledge made of ship-timber—Thick fog—Wolves—Walk along the shore—Remains of musk-cattle and reindeer—Nature of the coast—Danger from the ice—Irregular rise of the tide—Deer on the ice—Fruitless efforts to proceed northward—Cross over to Melville Peninsula—Gale—Again stopped by the ice—Dangerous position of the boat—Return to starting point—Meeting with our Esquimaux friends at Salt Lake—Deer begun to migrate southward—Walk across the isthmus to Repulse Bay.

The morning of the 26th was fine, with a fresh breeze from W.N.W. A visit which I had intended paying to the ladies was anticipated by their coming over to our side of the river, bag and baggage. They were accompanied by a very old man named Shad-kow-doo-yak, who was extremely infirm, being obliged to move about in an almost horizontal posture, supported by a stick. There were six women, (three old, the other three young,) the whole of them married. One of the latter appeared quite like a girl of ten years, and was rather good-looking, having more regular features, and being cleaner and more neat in her dress than the others. They were all tatooed on the face, the form on each being nearly the same, viz. a number of curved lines drawn from between the eyebrows up over the forehead, two lines across the cheek from near the nose towards the ear, and a number of diverging curved lines from the lower lip towards the chin and lower jaw. Their hands and arms were much tatooed from the tip of the finger to the shoulder. Their hair was collected in two large bunches, one on each side of the head; and a piece of stick about ten inches long and half-an-inch thick being placed among it, a strip of different coloured deer-skin is wound round it in a spiral form, producing far from an unpleasing effect. They all had ivory combs of their own manufacture, and deer-skin clothes with the hair inwards; the only difference between their dresses and those of the men being that the coats of the former had much larger hoods, (which are used for carrying children,) in having a flap before as well as behind, and also in the greater capacity of their boots, which come high above the knee, and are kept up by being fastened to the girdle. Some needles, beads, and other trifles were given them, at which they manifested their joy with loud shouts and yells, differing from the men in this respect, who received what was given them in silence, although they were evidently much pleased.

In the forenoon we were joined by two fine-looking young fellows who had just returned from hunting deer, in which they had been successful, having driven a large buck off one of the islands into the water and speared it there.

One of the women had been on board the Fury and Hecla, both at Igloolik and Winter Island, and still wore round her wrist some beads which she had obtained from these vessels. This party consisted of twenty-six individuals, there being four families.

All the cargo being placed in security and the Magnet well moored in our little land-locked harbour, the party, assisted by four Esquimaux, commenced dragging the North Pole up the stream.

The latitude of our landing place was found to be 66° 32' 1" N., being about seven miles further south than it has been laid down on the charts. The variation of the compass by an azimuth was 58° 37' 30" W. This I afterwards found to be erroneous, probably arising from local attraction. The rate of the chronometer had become so irregular that it could not be depended upon for finding the longitude, and during the winter it stopped altogether.

When about to put on a pair of Esquimaux boots, one of our female visitors, noticing that the leather of the foot was rather hard, took them out of my hands and began chewing them with her strong teeth. This is the mode in which they prepare and soften the seal skin for their boots, and they are seldom without a piece of leather to gnaw when they have no better occupation for their teeth. At half-past nine P.M. the men returned from the boat, having been absent since half-past seven in the morning. They had with much labour dragged her three miles through a succession of rapids, the channel being so obstructed with large boulder stones and rocks, that the most of the party were obliged to be almost continually up to the waist in ice-cold water. The boat had received some severe blows and rubs, but no material damage. The worst part of the river had been passed, and it was only a mile and a-half farther to the lake (named by the Esquimaux Chi-gi-uwik) from which it takes its rise. The Esquimaux who had assisted us were paid with a large knife each.

Two nets that had been set produced four salmon, but the best season for catching these fish was over, as they had now returned to deep water. The evening was cloudy with a strong and chilly breeze from N.N.W. Temperature of the air at 10 P.M. 35°.

27th.—As soon as the men had finished breakfast they carried each a load over the rocks to where the boat lay.

I this morning tried some of our male friends with a little tea and biscuit, which they did not relish nearly so well as the ladies had done the previous evening. Indeed, one of the latter, whom I have already mentioned, knew what biscuit was the moment she saw it, and said she had eaten some when on board Captain Parry's ships. I remained at our landing-place until the afternoon to obtain some observations. That for latitude gave a result different only 4" from that of yesterday. Having engaged three Esquimaux to carry up some things that were still to be taken, at one o'clock I followed my men and came up with them some distance up the lake. As we could not prevail on any of the Esquimaux to accompany us as guides, they left us here, and I sent back John Folster and Ouligbuck to take care of the property left behind.

Our course was nearly N.N.W., but a gale of head wind impeded our progress greatly. The temperature of the air was 52°; water of lake 40°. A few hours' poling, pulling, and tracking brought us to the end of the lake, which is about six miles long, from two hundred yards to half a mile broad, and in some places thirty fathoms deep. The lake, as well as the stream up which we had come, was named after our boat. We now turned to the westward and entered a narrow passage one-and-a-half miles long, which connects the lake we had passed through with the next one; the current was strong, but between poling and tracking we soon got into still water. Our course now turned again to the N.N.W., and after proceeding a mile in this direction, we put on shore for the night in a small bay, where we found a good supply of a plant (*andromeda tetragona*), which answers very well for fuel.

28th.—We did not get under weigh this morning until 6 A.M. as the men had a hard day's work yesterday, and did not get to rest until a late hour. The lake continues to trend in the same direction as before, but the banks are neither so high nor so rocky, being covered with short grass in many places instead of moss. The wind still kept ahead, so that it was past ten in the morning before we arrived at a portage, and while two of the men were preparing breakfast, the others were employed carrying over some of the baggage. This portage, which I named after Flett, one of the steersmen, was half a mile long; and being in some places soft and in others stony, it was half-past four before we were afloat in the lake on the other side of it. It being calm, great numbers of fish were seen in this small body of water, which was narrow and only two-and-a-half miles long, with a deep bay on each side, which gave it the form of a T. It received the name of Corrigal, after one of my men. We lost our way here for a short time, having entered a wrong arm of the lake. At 8 P.M. we arrived at another portage, which being a short one was soon got over. We pulled in a N.W. direction across this lake for about three miles to a shallow streamlet that flows from it; here we were to make our third and I hoped our last portage. We left this for our next morning's work, as it was now half-past 10 P.M. There was a great number of stones set up here for the purpose of frightening the deer into the water. A large white wolf was seen.

The morning of the 29th was raw and cold, with a gale of wind from N.W. by N. We got over the portage (which, although short, was covered with rough granite stones that stuck to our boat's iron-shod keel like glue) at 20 minutes after 6, and embarked on what I then supposed was another lake, but which afterwards turned out to be a portion of the second lake we had entered, and the largest body of fresh water we had yet seen. I named it after my much-respected and kind friend, Alexander Christie, Esq., Governor of Red River Colony, whose name has been so often favourably mentioned by Arctic travellers.

After pulling W.N.W. for eight miles, we were again in doubt about the route, and whilst on my way to some high ground in order to ascertain it, I shot a fine buck with an inch and a half of fat on his haunches.

We advanced two miles to the head of a small inlet, whence I set out with one of the men to a neighbouring rising ground to endeavour to obtain a view of our future route, and, if possible, to get a sight of the sea. After a fatiguing walk over hill and dale, our eyes were gladdened with a sight of what we so anxiously looked for, but the view was far from flattering to our hopes. The sea, or rather the ice on its surface, was seen apparently not more than twelve miles distant, bearing north; but there was not a pool of open water visible. It was evident that our detention in the lakes had as yet lost us nothing. Returning at 8 P.M., I sent four men in two parties to endeavour to discover the best route, one party being ordered to trace a considerable lake in a N.N.W. direction, and, if possible, discover its outlet.

30th.—The men sent off last night returned between 1 and 2 this morning: those who went to the N.W. reported that there was a small stream flowing towards the Arctic Sea from the farthest extremity of the lake they had traced.

As this account agreed with what we had heard from the Esquimaux, there was no doubt that we were now in the right track. We had to cross two portages, each a quarter of a mile, and traverse a lakelet one mile in extent, before we reached the body of water which the men had traced to its outlet. It was half-past 2 before we accomplished this work, there being many obstructions in the form of large granite stones, among and over which we had to drag the boat.

The lake in which we now found ourselves is upwards of 27 fathoms deep, about 6½ miles long, and not more than half a mile broad; it lies nearly N. by W., and is bounded by banks much more steep and rugged than any we had yet passed, being in some places two or three hundred feet high. It is situated in latitude 66° 55' N., and longitude 87° 35' W. We found that the longest and most difficult portage was now before us. By the time we had the baggage carried half way over it was getting late, and we did not take dinner until 9 P.M.

The following morning was cloudy, with a cold north breeze, which was not at all unfavourable for the work we had to do. We went to work at an early hour, but our advance was very slow, as the portage fully realised the bad opinion that we had formed of it. Hitherto, by laying the anchor out some distance ahead, and having a block attached to the bow of the boat by a strop, or what sailors call a swifter, passing round her, we could form a purchase sufficiently strong to move her with facility, but here our utmost exertions were required, and the tracking line was frequently broken. A piece of iron an eighth of an inch thick, which lined the keel from stem to stern, was actually drawn out and doubled up, so that it was necessary to remove the whole. At half-past 10, when half-way across, we breakfasted, after which we met with a bank of snow, over which we went at a great rate. The latitude, 66° 59' 37" N., was observed. Near the extremity of the portage there were some ponds of water deep enough to float the boat, that helped us not a little. The descent of a steep bank fully a hundred feet high brought us into another fine lake eight miles long and one mile broad, lying nearly north and south, with steep rocky shores on its west side: the place where we came upon its waters was about three miles from its southern extremity. This lake was named "Miles," after a friend. As it was quite calm, we pulled up due north and entered a narrow inlet, out of which there was no passage. We had passed at a mile and a half from this a stream flowing from the lake, but it looked so insignificant that I could not suppose it to be the same that the Esquimaux had reported as having sufficient water for floating the boat. It was now too late, however, to look for any other exit, and we all betook ourselves to rest after a hearty supper, for which the fatigues of the day gave us an excellent appetite. Some of the men had large pieces of the skin stripped from their backs whilst lifting the boat over the various obstructions on the portage.

1st August.—Finding that there was no likelihood of there being any other outlet to the lake than the one we had seen, we took out the cargo, and hauling our boat over a shallow part, we reloaded and soon entered a narrow lake, the waters of which were very muddy. At half an hour before noon we landed to have breakfast, and the latitude 67° 4' 22" N., variation of the compass 66° 38' W., were observed. The shores of this lake, being covered with a rich pasturage and a great variety of flowers, afforded a pleasing contrast to the country we had hitherto travelled through. There were great numbers of marmots here, with a well-beaten path leading from one burrow to another. After dragging the boat over many shallows, we arrived a little after 5 P.M. at high-water mark, in latitude 67° 13' N., longitude 87° 30' W. The tide being out, and there not being sufficient water to float the boat, I decided on remaining here until the flood made.

The recent foot-tracks of two Esquimaux were seen on the sand.

A short distance below where we stopped, the stream we had descended empties its waters into a small river which flows from the westward.

2d.—As the tide did not rise so high by two feet during the night as it had done the previous day, the boat did not float; we were, consequently, obliged to carry our baggage a mile further down the stream, and afterwards, with much trouble, haul our boat over numerous shoals. We were now afloat in a salt-water lake, and on passing a small point two Esquimaux tents came in view. Not having got breakfast, I landed with the interpreter, and, whilst the men were cooking, went to ascertain if there were any inhabitants. All was quiet inside, but after calling once or twice outside the door of one of the tents, an old woman made her appearance, apparently just out of bed, as she was very coolly drawing on her capacious boots, whilst she surveyed her visitors without showing the slightest symptoms of alarm, although I afterwards learned that I was the first European she had ever seen. An old man soon after popped out his head alongside that of his better half, who appeared to be endowed with a flow of language which set all his efforts to say anything at defiance. A few trifling presents put us all, in a few minutes, on a most friendly footing. Their report of the state of the ice in the large bay before us was far from encouraging; they said that there was seldom sufficient water for the passage of one of their small canoes, and present appearances led me to suppose that they were correct. The name of the man was I-il-lak, of the woman Rei-lu-ak. The remainder of the party, consisting of their two sons and their wives, had gone a day's journey inland to hunt the musk-ox. From a chart drawn by the woman, who, as is usual, (at least among the Esquimaux) was much the more intelligent of the two, I was led to infer that there was no opening leading into the large bay but through the Strait of the Fury and Hecla, and Prince Regent's Inlet.

As soon as breakfast was over, in which our new friends joined us, we crossed the lake, which is 6 miles long by 1½ broad, and put on shore three of the men (W. Adamson, H. Mineau, and Nibitabo) who had assisted us across, and were now to walk back to Repulse Bay, a distance of forty-three miles. By them I sent orders to John Folster (the man left in charge) to make every possible preparation for wintering, and to keep up a friendly intercourse with the natives. My crew now consisted of George Flett, John Corrigal, Richard Turner, Edward Hutchison, Peter Matheson, Jacques St. Germain, and William Ouligbuck. We now passed for two miles through a narrow channel—not more than 40 yards wide—among pieces of ice which were carried along with great rapidity by the ebb tide that had just commenced; this led us into the deep inlet which we had seen on the 29th ult. This inlet I named after Donald Ross, Esq., Chief-Factor. We found but little open water; by keeping near the rocks, however, we made some progress northward by using our ice-poles, and after advancing a mile or two I went upon a piece of ice and obtained the latitude 67° 15' N. by a meridian observation of the sun in quicksilver. About eight miles to the north of this we passed a rocky point, which was named after Chief-Factor Hargrave, the gentleman in charge of York Factory when the expedition was fitted out, and who afforded every possible assistance towards its proper equipment. This point is formed of granite and gneiss, and has a very rugged appearance, there being neither moss nor grass on the rocks to soften their asperities.

At 7 A.M. on the 3rd, when a few miles past Point Hargrave, being completely stopped by ice, we put ashore and found a large wooden sledge, which we cut up for fuel. The wood was evidently the planks of some vessel (probably of the Fury or Sir John Ross's steamer the Victory) as there were holes in it bored with an auger. After working our way a mile or two further, we arrived at a high rocky cape having three elevations upon it lying east and west from each other. This headland, which was honoured with the name of the lady of Sir John H. Pelly, Bart., Governor of the Hudson's Bay Company, is situated in latitude 67° 28' N., longitude by account 87° 40' W.; variation of the compass 82° 36' W.

It was low water to-day at 11 A.M., the fall of the tide being 8½ feet, and the depth of water within a hundred yards of the beach from 3 to 5 fathoms, on a bottom of mud or sand.

Shortly after noon a fog came on so thick that we could only see a few yards round us; we, however, pushed our way for 2½ miles beyond Cape Lady Pelly, along a flat coast lined with mud banks from eight to ten feet high, frozen solid within a foot of the surface. At 4 P.M. the ice was too closely packed to allow us to proceed; we therefore turned towards the shore, and after some trouble effected a landing. The fog still continued so thick, that, after wandering about for a few miles, I had much difficulty in finding the boat again, hid as it was by the surrounding masses of ice. We were much at a loss for drinkable water, there not being a drop in the neighbourhood but what resembled chocolate in appearance.

In the forenoon some wolves, part of a band that had serenaded us last night with their dismal howlings, were seen prowling about; and a white-winged silvery gull (*L. leucopterus*), a diminutive sandpiper (*tringa minuta*), and a marmot were shot.

4th.—There was a drizzling rain with thick fog all night, but not a breath of wind. As the tide flowed the ice moved slowly and silently round us, so that in the morning we had not more than a yard or two of open water near us, being blocked in on all sides by pieces from 15 to 20 feet thick. The rise of the tide was not less than nine feet. In the forenoon I walked upwards of five miles along the shore to the northwestward, passing a few low sandy points about a mile and a half from each other, which formed a succession of small bays, into each of which a ravine with high and steep mud banks opened, down which a streamlet of pea-soup-coloured water flowed. We fell in with the heads and horns of several musk cattle and reindeer, and saw recent footmarks of some of the latter, but they had probably been driven some distance away by the wolves we saw yesterday. Marmots were numerous in every direction, chattering to each other, and rising on their hind legs to obtain a better view of the strangers. Many golden plovers and different kinds of sandpipers were flying about, and a jager (*L. parasiticus*) was shot: some plants were also collected. The travelling along this coast was extremely fatiguing, being very often nearly knee-deep in a very adhesive mud.

The thermometer rose as high as 70° in the forenoon; in the afternoon it fell to 48°; and in the evening the weather was cold and unpleasant, with heavy rain.

5th.—During the greater part of last night the rain continued, but it was perfectly calm, although by the lead of the clouds we were in hopes of a breeze of wind off shore. Our boat being in danger of injury from some heavy masses of ice that were turning over near us, we moved a dozen yards nearer the land. Our new situation, however, was little better than the one we had left, for as soon as the tide began to ebb large pieces of our "enemy" broke away and fell with a loud crash close alongside of us. It was high water this morning at 3 o'clock, the rise of the tide being 11 feet 6 inches, whilst that of yesterday evening was only 5½ feet, an irregularity resembling that which was observed by Captain Sir J. Ross on the shores of Boothia. The temperature of the air in the morning was 46°, but rose to 65° during the day, which was very hazy, with occasional showers and a fresh breeze offshore; but this had not the slightest effect upon the ice, and led me to believe that the Esquimaux report as to the navigation being always obstructed here is correct.

Seeing that there was no probability of our getting along shore towards Dease and Simpson's farthest, I determined to retrace our route, and if possible cross over to Melville Peninsula for the purpose of surveying its western shore, towards the Strait of the Fury and Hecla.

In the evening, when the tide, which on the present occasion rose only 4½ feet, was in, we endeavoured to extricate ourselves; and after some hours of hard labour in chopping off some points of ice, and pushing aside such pieces as were not aground, we got a few hundred yards from the beach, and into water a little more open.

About half-past ten a young buck was observed on a piece of ice half-a-mile to seaward, having been forced to take the water to avoid some wolves, one or two of which were seen skulking along shore watching for the return of the animal. The state of our larder did not permit us to be merciful, so the poor deer had little chance of escape from his biped and quadruped enemies when acting in concert. After a long chase he was shot whilst swimming from one floe to another. Having pulled and poled along shore all night, we landed for breakfast at 8h. 30m. A.M., on the 6th, about three miles to the south of Point Hargrave. The continued rain and fog had so completely saturated everything with damp that we had not a dry stitch of clothes to put on, and our bedding and fuel were in the same state; fortunately the weather was mild, so that we did not feel much inconvenience from this.

Finding that the ice was clearing away a little—the effect of a south-east wind,—we directed our course towards the nearest point of Melville Peninsula, which bore east (true) of us, distant ten miles, and after threading our way among much heavy and close-packed floes, which obliged us to make frequent and long detours, after five hours' hard work we reached the land during a thunder-storm accompanied by torrents of rain.

Our landing place was a long rocky point having a deep ice-filled inlet on its south side. To this point I gave the name of Cape Thomas Simpson, after the late enterprising traveller of that name.

As we could not proceed on account of the thick fog and the state of the ice, we secured the boat to the rocks, and the men although drenched to the skin went immediately to sleep, eighteen hours of hard work at the oars and ice-poles having thoroughly tired them all.

During the night of the 6th the weather was thick with occasional rain, but about 6 in the morning of the 7th a fresh breeze from the south-east dispersed the fog. As soon as it was cleared up we renewed our voyage, but our progress was very slow, having our old opponent to contend with; in four hours we gained as many miles and were again stopped. Seeing some deer near the beach, we landed, and whilst two of us had a fruitless chase after them the remainder of the party were busy cooking and drying our clothes, blankets, &c. The temperature of the air was 52°, that of the water 35°.

The breeze gradually increased as the day advanced, and went round to the east, which drove the ice a short distance from the shore. We embarked again between 9 and 10 A.M., and ran to the eastward for a league or more, when the breeze having changed into a heavy gale, our boat ran great risk of being injured by the ice, of which we found it impossible to keep altogether clear. We therefore pulled up to a number of grounded pieces (a line of which completely barred us from the shore), and made fast to the largest of them. In getting this far we were in much danger from the falling, or breaking off, of overhanging masses (some of them 20 feet in height), which were crashing all around us, and under which we had frequently to pass. At 5 A.M. our floe got afloat, and began driving to leeward at a great rate. We just got the boat clear in time to prevent its being crushed against a berg that still remained fast. Some of the smaller pieces lying between us and the land having now floated, we managed to clear a passage for ourselves; yet although we had only a quarter of a mile to go, so strong was the gale that it required the utmost exertions of six men at the oars to reach the shore, when, having secured the boat and raised an oilcloth to keep off the rain, which had again commenced, we had our supper of pemmican and water, and retired to bed for the night.

8th.—On getting up this morning I found that it had become quite calm, and that the ice was coming in so thick and fast with the flood tide, that we had to move from our position as fast as possible. On pushing out to sea it soon became apparent that we could not proceed on our course, and that there was but little open water in the direction from whence we had come, and even that was fast filling up. As we could neither advance nor remain in safety where we were, there was only one course open to us, and that was to return towards the place from which we had started.

It was now evident that this large bay was completely full of ice; for had this not been the case, the gale of yesterday must have cleared the coast for many miles. It was with a sad heart that I turned the head of our boat towards our starting point, where I purposed to await some favourable change in the state of the ice, and at the same time learn how the people left at Repulse Bay were getting on with preparations for wintering, which now appeared inevitable. The weather continued so much overcast that no observations could be obtained. In the afternoon a light breeze sprang up from W.N.W., which enabled us to reach in a short time Ross Inlet, where we had some trouble in finding the entrance of the river on account of the altered appearance of the rocks, it being now nearly low water and the shore clear of ice, compared with what it formerly was. We had much difficulty in towing up to the Salt Lake before mentioned, as the narrow but deep channel which led to it was, at this state of the tide, one continued rapid, and so strong was the force of the stream that our tracking line broke. We were soon snug in the Salt Lake, but had not been more than half an hour under shelter before almost every spot of open water outside was filled with ice, so rapidly had it followed in our wake.

When we arrived opposite the tents of our Esquimaux friends, they came running down to the beach led on by the old lady whose fluency of speech I have already remarked, and who appeared determined to sustain her character on this occasion by making more noise than all the others put together, and expressing her joy at our return by loud shouts. The old people had during our absence been joined by the musk-ox hunters, two fine young active-looking fellows (named Ark-shuk and I-vit-chuck) and their wives. These women were the cleanest and best-looking I had yet seen. They were tatooed much in the same way as those at Repulse Bay. The hunters said they had been unsuccessful, but as each of the women had the tail, or a portion of the shaggy hair of the neck, of a musk-ox in her hand as a musquito flapper, their veracity was rather doubted. There was only one child with them, a sickly-looking boy of six or seven years, stepson to a man named Shi-shak, who arrived about an hour after us in his kayak from an unprofitable walrus hunt.

I learnt from our Esquimaux acquaintances that the deer had commenced migrating southward. This being the case, I prepared to walk across to Repulse Bay to see what progress the party left there had made in their work. The weather had been so cloudy for the last week that no observations of any value could be obtained.

Leaving three men and Ouligbuck's son in charge of the boat, I started at 6.30 A.M. on the 9th, in company with Corrigal, N. Germain, and Matheson, to cross the isthmus, taking a S.S.E. direction; but it was impossible to keep this course for any great distance, as we were forced to make long circuits to avoid precipices and arms of lakes. After a most fatiguing day's march over hill and dale, through swamp and stream, we halted at half-past 6 P.M. close to the second portage crossed on our outward route. To gain a distance of twenty miles we had travelled not less than thirty. Our supper was soon finished, as it was neither luxurious nor required much cooking, consisting of our staple commodities pemmican "cold with water."

10th.—The morning was raw and cold with some hoar frost, and there not being a blanket among the party and only two coats, our sleep was neither long, sound, nor refreshing. In fact I had carried no coat with me except a thin Macintosh, which, being damp from the rain of yesterday, had become an excellent conductor of caloric, and added to the chilly feeling instead of keeping it off.

There is one advantage in an uncomfortable bed, it induces early rising, and it proved so in the present instance, for we had finished breakfast and resumed our journey by half-past 2 A.M. The travelling was as difficult as that of yesterday, but we had the advantage of a cool morning and got on more easily. At 7 o'clock we arrived at the narrows which separate Christie and North Pole Lakes, where we found the greater number of the Esquimaux we had seen, encamped, waiting for deer crossing over. Some of them immediately got into their kayaks and paddled across to our side of the lake, but with so much caution that it was evident we had not yet wholly gained their confidence.

At 2 P.M. we arrived at Repulse Bay with most enviable appetites, but rather foot-sore, our shoes and socks having been entirely worn through long before we reached our destination.

CHAPTER FOUR

State of things at Repulse Ray—Determine to discontinue the survey till the spring—Reasons—Party sent to bring over the boat—Fix on a site for winter residence—Ptarmigan—Laughing geese—Eider and king ducks—Visits of natives too frequent—Return of the party sent for the boat—Report the bay more closely packed than before—Preparations for wintering—Fort Hope built—Proceed to North Pole and Christie Lakes to look out for fishing stations—Purchase dogs—Wariness of the deer—Flocks of geese pass southward—Blue-winged and snowgeese—Their habits—Snow-storm—Its effects—Return to Fort Hope—Daily routine—Signs of winter—Deer numerous—Quantity of game killed—Provision-store built of snow—Great fall of snow—Effects of the cold—Adventure with a deer—Visited by a party of natives—Their report of the ice westward of Melville Peninsula—An island said to be wooded—Produce of the chace in October—Temperature—Two observatories built of snow—Band of wolves—A party caught in a snow-storm—Esquimaux theory of the heavenly bodies—Temperature of November—Diminished supply of provisions.

On our arrival at Repulse Bay we found the men all well, but getting no more fish and venison than was barely sufficient to support them. Having taken but a scanty breakfast, I fully enjoyed my dinner here, but I reversed the usual order of eating the same, taking my venison steak first (it being soonest cooked), and salmon as second course.

This was to me the most anxious period during the expedition; nor will this appear strange when I mention that it was necessary to decide, and that promptly, on one of two modes of proceeding, namely, whether to leave the whole survey to be completed during the following spring and summer, or to endeavour to follow it up this autumn.

After mature consideration I determined on adopting the first of these measures, and giving up all hopes of prosecuting the survey at present.

My reasons for arriving at this conclusion I shall briefly mention, as such a step may appear rather premature. I saw from the state of the ice and the prevalence of northerly winds that there was no probability of completing the whole of the proposed survey this season; and although part of the coast, either towards the Strait of Fury and Hecla, or towards Dease and Simpson's farthest, might be traced, yet to accomplish even this might detain us so long that there would be no time to make the necessary preparations for wintering, and we should thus be under the necessity of returning to Churchill without accomplishing the object of the expedition, or, if we remained at Repulse Bay, run the risk of starving, for I could obtain no promise of supplies from the natives, and all the provisions we had carried with us would not go far to support the party throughout the winter. We should thus have to depend almost, if not altogether, upon our own exertions for the means of existence both in regard to food and fuel.

It ought to be borne in mind that we were differently situated from any party that had hitherto gone to these cold and barren regions. The resources of the country were quite unknown to us; it was not likely that the deer would remain near at hand all winter, as we were at too great a distance from the woods, and it was very evident, for the same reason, that we should not be able to procure any sort of fuel after the first fall of snow, which there was little doubt would occur some time in September.

Before reaching the Arctic Sea to the west of Melville Peninsula, I was for various reasons inclined to agree with the opinion of Sir John Ross, "that Boothia was part of the continent of America." This opinion was strengthened when I observed the great rise and fall of the tide, which must have affected the tides at the Castor and Pollux River, had there been a strait of any width separating Boothia from the mainland, unless indeed the assumption of Captain Sir J. Ross, that "the sea to the west of Boothia stands at a higher level than that on the east side," be correct. In that case there would be a continual easterly current, which could scarcely fail to have been noticed by so acute an observer as Simpson.

Retaining one man with myself to guard our stores and attend the nets, on the 11th I sent over the remaining six to assist in bringing over the boat. Ouligbuck had now been about two days looking for deer, and I began to feel anxious about him, when he made his appearance between 9 and 10 A.M. with the venison of a young deer on his back.

As soon as my companion had returned from the nets, out of which he got no fish, I took a walk for the purpose of looking out for fishing stations and a site for our winter house. For the latter I could find no better place than a narrow but not deep valley within a few hundred yards of our landing-place, and about a hundred and fifty from North Pole River on its east side. There appeared to be various small bays along shore to the eastward which were likely to produce fish. A flock of laughing geese (*anser albifrons*) flew past quite close to me; but having only my rifle, I could but send a ball after them and missed as was to be expected.

In a small pond an eider-duck was observed with her young brood apparently not more than twelve days old. The male eider and king ducks had already left this quarter, having migrated to the southward.

12th.—A cloudy day with a strong breeze from N.N.W. Two salmon and a trout were got from the nets, but Ouligbuck killed no deer. In the evening, when on my way to set a net in a lake at no great distance, I fell in with a covey of ptarmigan, (*T. rupestris,*) most of the young being strong on the wing, and bagged eighteen brace in an hour or two. Knocking down those birds on this day made me half fancy myself among the grouse in my own barren native hills.

On the 13th the weather was raw and cold with frequent showers, and a gale of wind from the same quarter as the day before. Four salmon were caught, and a deer was shot. The thermometer varied from 36° to 38°. Four Esquimaux men and two women visited us to-day.

The 14th was much like the 13th, but there was no rain. As the visits of the natives had now become rather frequent, and as they brought nothing with them, but appeared to expect both food and presents, I bade Ouligbuck say that we could not afford to feed them any longer, and that they had better return to their huts, where I knew they were killing deer enough to support themselves. On returning from my daily walk, I found that our friends had taken leave rather hurriedly, having been detected appropriating some salt fish, which they could not eat. For this they were sharply reprimanded by the interpreter, and one of the ladies was most ungallantly accused by her husband of being the offender. Corrigal and I hauled the seine in the evening and caught thirty-three salmon; fourteen more were got out of the nets.

15th.—This was a beautiful day throughout. In the evening, the sky being clear and cloudless, some stars were visible, and a few streaks of orange-coloured aurora showed themselves to the southward. The seine was again hauled, and thirty-two salmon (some of them very small) caught, whilst the nets produced eleven more. Just as we were landing our fish, the men who had been taking over the boat made their appearance, being a day earlier than I expected. By keeping the proper route three of the portages were avoided, and they had the advantage of a fine fair breeze all through the lakes. The large bay (Akkoolee) was reported as being more closely packed with ice than before. This was nothing but what I should have expected after the late north-westerly winds.

The two Esquimaux, Arkshuk and Ivitchuck, ("Anglice" Aurora and Walrus,) who had been engaged to aid in dragging the boat over the portages, had wrought well, and readily accommodated themselves to the habits of the men. They were well recompensed, and Ivitchuk (a merry little fellow) was engaged to accompany me on my intended spring journeys.

The boat was for the present left at North Pole Lake, as it might still be required there.

The 16th was a day of rest, and the 17th was so stormy and wet that little work could be done.

All hands were now busily employed making preparations for a long and dreary winter; for this purpose four men were set to work to collect stones for building a house, whilst the others were occupied in setting nets, hunting deer, and gathering fuel. Our work was much impeded by rainy weather, particularly the house building, as the clay or mud was washed away as soon as applied.

We found that our nets were so much cut up by a small marine insect from a half to three-quarters of an inch long, resembling a shrimp in miniature—the favourite food of the salmon—that it was quite impossible to keep them in repair. I thought to destroy their taste for hemp by steeping the nets in a strong decoction of tobacco, but it had no effect.

On the 2d September our house was finished; its internal dimensions were 20 feet long by 14 feet broad, height in front 7½ feet, sloping to 5½ at the back. We formed a very good roof by using the oars and masts of our boats as rafters, and covering them with oilcloth and moose skin, the latter being fixed to the lower or inside of the rafters, whilst the former was placed on the outside to run off the rain. The door was made of parchment deer-skins stretched over a frame of wood. The walls were fully two feet thick, with three small openings, in which a like number of windows, each having two panes of glass, were placed.

Our establishment was dignified with the name of Fort Hope, and was situated in 66° 32' 16" N., longitude (by a number of sets of lunar distances with objects on both sides of the moon) 86° 55' 51" W. The variation of the compass on 30th August was 62° 50' 30" W.; mean dip of the needle, and the mean twice of a hundred vertical vibrations in the lide of declination 226".

A sort of room was formed at one end by putting up a partition of oilcloth. In this, besides its serving as my quarters, all our pemmican and some of the other stores were stowed away.

From the 5th to the 13th I was up at North Pole and Christie Lakes in the boat with three men, our object being to look out for fishing stations, and also to purchase dogs from the Esquimaux. The wind being from the north, we did not reach the Esquimaux encampment till the 10th. They had shifted their tents from the narrows to a small point about eight miles up Christie Lake, where the deer were more numerous, among which they seemed to have made great havoc, to judge by the abundance of skins and venison lying in all directions. Our friends were delighted to see us, and had improved much in appearance, the only poor animals about them being their dogs, which appeared to get no more to eat than was barely sufficient to keep them in life. I looked out four of the best, being all I wanted at present, for which I promised a dagger each, intending to take them with us on our return. During our stay here a band of deer came to the edge of the lake, and after feeding a short time took the water. Three of the natives slipped noiselessly into their kayaks, and lay waiting, until the deer were far enough out in the water, to intercept them, but just as they were on the eve of starting the wind changed a little, and the deer smelling their enemy wheeled about, and were soon in safety on the beach from which they had started.

Many large flocks of Hutchins and snow geese had been seen for the last few days passing to the southward. The blue-winged goose of Edwards is by some ornithologists considered as the young of the last named bird in one of its stages towards maturity, but this opinion I believe to be erroneous, for the following reasons.

During a ten years' residence at Moose Factory, on the shores of Hudson's Bay, I had many opportunities, every spring and autumn, of observing both the snow and the blue-winged goose in their passage to and from their breeding places, the marshes near Moose being favourite feeding ground.

In spring both species are very nearly alike in size, the blue-winged goose, although shorter, being rather the heavier bird. In the autumn there are four distinct varieties, two of which exactly resemble in size and plumage those seen in the spring, whilst the others are much smaller, and differ much from these and from each other in their markings; the young of the snow goose being of a light grey colour, darkest on the head and upper part of the neck; whilst the young of the blue-winged goose is of a dark slate colour, approaching to black on the head and neck. Neither do the young separate from the old, as has been asserted; for families may be seen feeding by themselves all over the marshes, the old bird keeping a sharp look-out, and giving timely warning to her brood of any approaching danger. In fact the Indian, who has thoroughly studied the habits of the bird, takes advantage of her affection for her young, and of their attachment to their parent, to make both his prey. Well knowing that the young are easily decoyed by imitating their call and by mock geese set up in the marsh, and that the old bird, although more shy, will follow them, he waits patiently until she comes within range; if he shoots her he is pretty sure to kill the greater part of the others, as they continue to fly over and around the place for some time after.

During the night of the 10th, when near the north end of the lake, we experienced one of the severest snow storms I ever witnessed. As we were sleeping on shore we never thought of putting up any sort of shelter; the consequence was that in the morning we were covered with snow to the depth of a foot. Our boat, which had been hauled up on the beach, was blown away from her fastenings, and carried several hundred yards into the lake among some stones. Being the only one of the party provided on the spot with Macintosh boots, it fell to my lot to wade out to the boat, throw overboard the ballast, lift her bows over the stones, and take a line to the shore; which, from having miscalculated the depth of the water, I found a more disagreeable task than I had expected. Fortunately the boat sustained no injury. It was now about 6 o'clock in the morning of the 11th, and as the storm continued unabated we made a sort of tent of our sails. In doing this the men got so wet and cold, from the snow thawing on them, that they could not even light their pipes.

In the afternoon the weather improved, and we were able to scrape a little fuel together, with which we cooked some salmon and boiled a kettle of tea, which made us feel quite comfortable again. We thus combined breakfast, dinner, and supper in one meal.

The hares had already acquired their winter coat, and the golden plovers and sandpipers had all disappeared, but some Lapland and snow-buntings and the shore-lark were still to be seen.

A little after noon on the 13th the wind shifted to the S.W., and we got under weigh to return home. A couple of hours brought us to the Esquimaux, where we stopped to take on board our dogs. A young lad also came with us to carry some medicine for the patriarch of the tribe, who was labouring under various complaints peculiar to old age. We arrived at North Pole River at 6 P.M., having had a beautiful run all the way.

As we were not likely to require the boat on the lakes again this season, she was hauled up and placed in security for the winter. While at the lake we had not been able to procure much more food than was necessary for our own use, but this may in part have been attributable to the bad weather.

The storm of the 10th had been much felt at our house, and so great was its force that the boat left there was lifted a few yards by it, but received no injury. Much heavy ice was driven into the bay and lay heaped up all along the shore.

Our house was still far from comfortable, the clay being quite wet and producing a most unpleasant feeling of dampness,—far more disagreeable than a much lower temperature with dry weather.

Our time was now continually occupied in collecting fuel, (portions of which, as soon as it became dry, were built up into small heaps on the rocks near the house,) in fishing, and in shooting deer and partridges.

The routine of our day's work was as follows: in the morning we were up before day-light; the men got their orders for the several duties they had to perform, which were principally carried on out of doors, and at which they set to work immediately after rolling up their bedding and taking breakfast. This meal usually consisted of boiled venison, the water with which it was cooked being converted into a very excellent soup by the addition of some deer's blood, and a handful or two of flour.

Our dinner, or rather supper, consisted of the same materials as our breakfast, and was taken about 4 or 5 o'clock; after that, my time was employed in writing my journal or making calculations; whilst the men were busy improving themselves in reading, arithmetic, &c., in which I assisted them as much as my time would permit. Divine service was read every Sunday when practicable.

On the 20th the pools of water were covered with ice sufficiently strong to be walked upon, and on the 28th some hooks were set under the ice on the lakes for trout. During the latter part of the month deer were very numerous. As many as seventeen were shot on the 28th, and on the following day ten more were got, seven of which were killed by myself within a few miles of the house. On the 29th a considerable portion of the bay was frozen over, and the seals were seen popping up their heads every now and then through the ice to keep breathing places open.

The weather during this month having been very changeable and stormy, and unfavourable for observations of all kinds, the sextant had frequently been exchanged for the rifle—a not unwelcome exchange to one addicted to field-sports "from his youth upwards."

Our sporting book for the month showed that we had been doing something towards laying in a stock of provisions for winter; 63 deer, 5 hares, one seal, 172 partridges, and 116 salmon and trout, had been brought in.

October.—During the first part of this month some of the men were employed in building a store of snow for our provisions, and covering it with two of the sails. On the 12th and three following days there was one continued storm which drifted the snow all round the house as high as the roof, and on the night of the 15th would have choked all our dogs that were chained outside, had not Adamson and another got up and cut their fastenings. On the 16th, when it cleared up, the thermometer first fell to zero.

The cold had now penetrated in-doors and frozen the clay on the walls, which made us much more comfortable. On attempting to open some books that had been lying on a shelf, I was surprised to find that the leaves were all frozen together; when I mention this, and also that our powder horns and every other article that was bound with brass or silver burst their fastenings, some idea may be formed of the dampness of our house whilst the clay on the walls was wet.

On the 19th, when out shooting, having killed one deer, I went in pursuit of another (a large buck) that had been wounded, and put four balls through him. Thinking that the last ball had settled the business, (for he had fallen,) I went carelessly up to him without re-loading my rifle, and when within a few yards I believe I apostrophized the animal much in the following strain—"Ah! poor fellow, you are done for at last!" when the deer, as if he had understood what I said, and thought I was adding insult to injury, sprung to his legs in a moment, and at a couple of bounds his horns were within a foot of me. Circumstanced as I was, I thought with Falstaff "that discretion was the better part of valour," and beat a hasty retreat, laughing heartily all the time at the strange figure we must have made. Taking the deer by the horns could have been of no use, and might have cost me some troublesome bruises and scratches.

Twelve Esquimaux and a boy visited us on the 23rd; among whom was the man (named Shi-ma-kuk) to whom the sledge belonged, part of which we had used for fuel when near Cape Lady Pelly with the boat. He was now rewarded, and apparently so much to his satisfaction that he would have had no objections to have another sledge burnt on the same terms. They reported that the bay, to the west of Melville Peninsula, had been packed full of ice ever since we were over there, until a few days before they came away, when there was some open water to be seen. Besides purchasing five dozen rein-deer tongues, a seal-skin full of oil, and some other articles, we added two good dogs to our team.

Among other information they told me that there was an island in Akkoolee (the large bay west of Melville Peninsula,) named Sha-took, (which means low or flat,) on which large trees grew; but they acknowledged that none of them had ever been on the island, although they had been near enough to see the trees distinctly. In this I believe their imaginations had deceived them, aided perhaps in some degree by a peculiar state of the atmosphere, during which the appearance of the land has been so distorted that it has been mistaken for woods. Some round sticks, probably spars belonging to one of the two vessels left in Prince Regent's Inlet, having been picked up along the west shore of Melville Peninsula, had no doubt strengthened the opinion they had formed. Two of their party whom we had never seen, were drowned in

Miles Lake by falling through the ice; the one in chasing a deer, and the other, it is supposed, in attempting to save his companion.

Our visitors left us on the 25th, promising to return soon with some deer-skin dresses. During the whole of the month we were occupied much the same way as in the previous one. Deer were numerous during the first part of it, but scarce latterly; sixty-nine were shot, but the produce of our nets had fallen very low, eighteen salmon and four trout being all we caught. The highest temperature of the month was 38°, whilst the lowest was 15°. Although there was a great deal of very stormy weather, there were some clear calm nights, of which I took advantage to obtain lunar distances.

Two observatories had been built of snow, with a pillar of ice in each (at the suggestion of Captain Lefroy, R.A.), the one for the dip circle, the other for an horizontally suspended needle to try the effects of the aurora upon it.

So much snow had fallen that it lay four feet deep on the roof of our meat store, and was near breaking the masts which supported it; so that we were obliged to raise its walls about a fathom to prevent such an occurrence in future.

On the 4th November, when out looking for deer a little before day-light in the morning, I observed a band of animals coming over a rising ground at a quick pace directly towards me. I at first supposed them to be deer, but on a nearer approach they proved to be wolves, seventeen in number. They continued to advance at full speed until within forty yards, when they formed a sort of half circle to leeward. Hoping to send a ball through one of them, I knelt down and took what I thought a sure aim at a large fellow that was nearest; unfortunately it was not yet broad day-light, and the rascals all kept end on to me, so that the ball merely cut off a line of hair and a piece of skin from his side. They apparently did not expect to meet with such a reception, for after looking at me a second or two they trotted off, no doubt as much disappointed at not making a breakfast of me as I was at missing my aim. Had they come to close quarters (which they sometimes do when pressed hard for food) I had a large and strong knife which would have proved a very efficient weapon. On my way home I shot three hares.

On the 5th two partridges were shot which very much resembled the tetrao saliceti, but which I suppose to be the T. mutus. The parasitæ found on them differed from those usually found on the willow grouse.

We began during this month to find that we could not afford fuel to dry our clothes; I therefore adopted the plan that a celebrated miser took to warm his food, by taking them under the blankets with me at night, and drying them by the heat of the body. This, it may be supposed, was not very agreeable, particularly when the weather became colder, for the moisture froze during the day on the blankets, which sparkled with hoar frost when I went to bed.

In the afternoon of the 9th we had one of the most severe snow storms that had yet been experienced, and I was much alarmed at the non-arrival of four men who had gone in the morning to examine some nets and set others in North Pole Lake eight miles from the house. Guns were fired to attract the attention of the party, who made their appearance at half-past 8 P.M., when we had given up all hopes of seeing them until the following day. They had been upwards of eight hours in coming as many miles, and were like walking pillars of snow when they came in. The four dogs they had with them were still missing, having run off with the sled as soon as they smelt the house. On the following day they were found entangled with one another, and the sled stuck fast against some rocks. One or two of the dogs were completely covered up with snow, but all safe.

About 2 P.M. on the 25th, two Esquimaux men and a boy, named Arkshuk (Aurora Borealis), Took-oo-lak (the falling stick), and Chemik-tee (snuff), came to see us with deer-skin clothes, &c. for barter.

I had a good deal of conversation through the interpreters with Arkshuk; whom I found rather intelligent and communicative. It appears that the favourite food of these Esquimaux is musk-ox flesh; venison ranks next, and bear and walrus are preferred to seal and fish. Their theory regarding the sun and moon is rather peculiar. It is said that many years ago, not long after the creation of the world, there was a mighty conjuror (Esquimaux of course), who gained so much power that at last he raised himself up into the heavens, taking with him his sister (a beautiful girl) and a fire. To the latter he added great quantities of fuel, which thus formed the sun. For some time he and his sister lived in great harmony, but at last they disagreed, and he, in addition to maltreating the lady in many ways, at last scorched one side of her face. She had suffered patiently all sorts of indignities, but the spoiling of her beauty was not to be borne; she therefore ran away from him and formed the moon, and continues so until this day. Her brother is still in chase of her, but although he sometimes gets near, he will never overtake her. When it is new moon, the burnt side of the face is towards us; when full moon, the reverse is the case.

The stars are supposed to be the spirits of the dead Esquimaux that have fixed themselves in the heavens, and falling stars, or meteors, and the aurora borealis, are those spirits moving from one place to another whilst visiting their friends.

The highest, lowest, and mean temperature of November were respectively +28°, -25°, and +0.68°. Only twelve deer, nine hares, and a few partridges had been shot, whilst our nets produced about sixty fish, the greater part of which were small.

CHAPTER FIVE

Winter arrangements completed—Learn to build snow-houses—Christmas-day—North-pole River frozen to the bottom—1st January—Cheerfulness of the men—Furious snow-storm—Observatories blown down—Boat buried under the snow—Ouligbuck caught in the storm—Dog attacked by a wolf—Party of natives take up their residence near Fort Hope—Esquimaux mentioned by Sir John Ross known to them—Boat dug out of the snow—A runaway wife—Deer begin to migrate northward—A wolf-chase—First deer of the season shot—Difficulty of deer-hunting in spring—Dimensions of an Esquimaux canoe—Serious accident to Ouligbuck—A conjuror—Preparations for the journey northward—Temperature—Aurora Borealis.

During December we completed our various buildings, and formed passages under the snow, so that we could without exposure go to any of them. There were four houses, viz.: one for provisions, another for fuel, a third for oil, dog's meat, &c., and a fourth for the men's spare luggage, for which there was no room in the dwelling-house, and which had been stowed in the tents until it was found necessary to take them down.

Being desirous of requiring as little assistance from the Esquimaux as possible, I attempted to build a snow house after the native fashion, and succeeded tolerably well; finding that the process was not so difficult as I anticipated, after a few trials one or two of the men became very good masons. We had now no encouragement to move much about, as there was no game to be seen, and the weather was very unsettled, and consequently no more exercise was taken than was necessary to keep us in good health. In stormy weather, not being able to get out of doors, the men wrestled or played some game which called the muscles into action, and thus kept up the animal heat.

On the 21st, the sun's lower limb rose about double his diameter above a rising ground to the southward, on a level with Fort Hope. On the 23rd and 24th, whilst looking out some good venison for our Christmas dinner, we examined our stock of such provisions, and found that we had not enough to last us until the return of the deer in spring; fortunately we had still a good supply of pemmican left.

Christmas-day was passed very agreeably, but the weather was so stormy and cold that only a very short game at foot-ball could be played. Short as it was, however, it was sufficiently amusing, for our faces were every moment getting frost-bitten either in one place or another, so as to require the continual application of the hand; and the rubbing, running about, and kicking the ball all at the same time, produced a very ludicrous effect.

Our dinner was composed of excellent venison and a plum-pudding, with a moderate allowance of brandy punch to drink a health to absent friends.

For some time past, washing the face had been rather an unpleasant operation, as any water that got among the hair froze upon it immediately. This is mentioned by Sir George Back as having occurred once to him at Fort Reliance, in 1833. On the 28th, North Pole River got frozen to the bottom, so that we were forced to go to a lake to the S.W. of Beacon Hill, about half a mile distant, for water.

The 1st of January was as beautiful a day as we could have wished to begin the new year with. There was a light air of wind, and the temperature varied from -23° to -26°. After a most excellent breakfast of fat venison steaks, all the party were occupied for some hours with a spirited game at foot-ball, at which there was much fun, the snow being so hard and slippery that several pairs of heels might be seen in the air at the same time.

My dinner consisted of part of a hare and rein-deer tongue, with a currant pudding as second course. The men's mess was much like my own, except that they had venison instead of hare. A small supply of brandy was served out, and on the whole I do not believe that a more happy company could have been found in America, large as it is. 'Tis true that an agreeable companion to join me in a glass of punch, to drink a health to absent friends, to speak of by-gone times and speculate on the future, might have made the evening pass more pleasantly, yet I was far from unhappy. To hear the merry joke, the hearty laugh, and lively song among my men, was of itself a source of much pleasure.

On the 7th the tracks of a few deer were unexpectedly seen within a few miles of the house; and on the following day the thermometer showed a temperature of -47°, the lowest we experienced during the winter.

The 9th was a more disagreeable day than any we had yet had. A storm from the north with thick snow-drift, and a temperature of 72° below the freezing point, made it feel bitterly cold. Fortunately we had some days before made a house for our dogs, else they must have inevitably been frozen to death. Such was the force of the gale for two days that both observatories were completely demolished, and wherever the snow banks projected in the slightest degree above the surrounding level, they were worn away by the friction of the snow-drift as if cut with a knife.

The thermometer indoors varied from 29° to 40° below the freezing point, which would not have been unpleasant where there was a fire to warm the hands and feet, or even room to move about; but where there was neither the one nor the other, some few degrees more heat would have been preferable.

As we could not go for water we were forced to thaw snow, and take only one meal each day. My waistcoat after a week's wearing became so stiff from the condensation and freezing of my breath upon it, that I had much trouble to get it buttoned.

The gale did not subside until the 15th, when we were busily employed repairing the damages done by the wind and drift. As a great weight of snow had lodged upon our boat, we were afraid she might be injured by the pressure, and some of the men were employed to search for her, but there was some difference of opinion about her exact situation, and it was two days before she was found, after digging to the depth of eight feet.

A stick was set up at one end of the boat that there might be no difficulty in finding the place again.

One cause of discomfort to me was the great quantity of tobacco smoke in our low and confined house, it being sometimes so thick that no object could be seen at a couple of yards' distance. The whole party, with the exception of myself, were most inveterate smokers; indeed it was impossible to be awake for ten minutes during the night without hearing the sound of the flint and steel striking a light. Of course I might to a great extent have put a stop to this, but the poor fellows appeared to receive so much comfort from the use of the pipe, that it would have been cruelty to do so for the sake of saving myself a trifling inconvenience.

This month was so stormy that the most of our time when we could get out of doors was passed in clearing away the snow that drifted about our doors and over the house, and in rebuilding and repairing. The boat, and also the stick that had been set up as a mark, were completely covered over. On the 18th Ouligbuck had gone out to hunt, and did not return till the 25th, after I had given up all hopes of ever seeing him again in life. It appeared that he had visited the Esquimaux at Christie Lake for the purpose of speaking to them about not having kept their promise regarding some oil that they said they would bring to us, and which they had omitted to do. He had been caught by the storm of the 18th before he reached his friends, and was obliged to build a snow hut, in which he passed the night comfortably enough. On the following morning, when it cleared up a little, he found that he was not more than two hundred yards from his destination, which the thickness of the weather on the previous day had prevented him from seeing.

One of the dogs we had lent this party to aid in drawing some provisions to the coast had a narrow escape from a wolf. Having broken loose she set out on her return home, when she was attacked by the wolf, and treated much in the same way that Tam o' Shanter's mare was by Cutty Sark, for

"The wolf had caught her by the rump,
And left poor Surie scarce a stump."

On the last day of January some Esquimaux, who were to take up their quarters near us, arrived with part of their luggage and provisions, and built their snow house near the south side of Beacon Hill. This would have been the best situation for our establishment, as it was completely sheltered from the northerly gales, but we were too late in making the discovery.

I visited the Esquimaux on the 1st February, and found the old man, named Shishak, and his wife in their comfortable house, which was so warm that my waistcoat, which had been frozen quite stiff for some time past, actually thawed. It was not easy to learn any of the peculiarities of these people, as Ouligbuck was rather shy about describing their habits. Ouligbuck's son informed me that even in winter they strip off all their clothes before going to bed.

When taking a walk on the 3rd I passed near the Esquimaux, and found one of them repairing the runners of his sledge. The substance used was a mixture of moss chopped up fine, and snow soaked in water, lumps of which are firmly pressed on the sledge with the bare hand, and smoothed over so as to have an even surface. The process occupied the man nearly an hour, during the whole of which time he did not put his hands in his mitts, nor did he appear to feel the cold much, although the temperature was 30° below zero.

On the 4th Ouligbuck set his gun for a wolf that had been prowling about for the last few days. The usual mode is to fix the gun to two sticks with its muzzle pointed to a bait placed at the distance of fifteen or twenty yards, with a line attached to it, the other end of which is fastened to the trigger; but Ouligbuck's plan was quite different from this. He enclosed the gun in a small snow house, in such a manner that there was nothing visible but the bait, which was not more than a foot from the muzzle, so that the shot could scarcely miss the head of the animal. When Ouligbuck went to his gun next morning, he saw the track of the wolf, and followed it to the dog-kennel, in which he had comfortably taken up his quarters; he immediately took the brute by the tail and dragged him outside much against his will, when he was soon dispatched with an ice-chisel. This animal was very large, but in the last stage of starvation, with a severe arrow or gun-shot wound in one thigh. He measured 5 feet 9 inches from the nose to the tip of the tail, (length of tail 1 foot 7 inches,) and his height at the shoulder was 2 feet 8 inches.

On the 7th a man named Ak-kee-ou-lik, who had promised us four seal-skins of oil, arrived and said that he could only let us have one, because the bears had broken into his "cache" and devoured nearly all its contents. This story I did not believe at the time, and I afterwards found out that it was false. I felt a good deal annoyed at the man's not keeping his promise, because we had depended much upon this supply for fuel and light. To save the former, we had during part of last month taken only one meal a-day, and discontinued the comfort of a cup of tea with our evening repast. Of oil, our stock was so small, that we had been forced to keep early and late hours, namely, lying occasionally fourteen hours in bed, as we found that to sit up in a house in which the temperature was some degrees below zero, without either light or fire, was not very pleasant. Fortunately we all enjoyed excellent health, and our few discomforts, instead of causing discontent, furnished us with subjects of merriment. For instance, Hutchison about this time had his knee frozen in bed, and I believe the poor fellow (who by-the-bye was the softest of the party) was afterwards very sorry for letting it be known, as he got so heartily laughed at for his effeminacy.

On the 9th, one of the Esquimaux women (wife of Keiktoo-oo) that came to see us, had a brass wheel 1⅓ or 2 inches in diameter fastened on her dress as an ornament. It was evidently part of some instrument, probably of some of those left by Sir John Ross at Victoria Harbour. I wished to purchase it, but she would not part with it.

15th.—Akkeeoulik brought over a large and heavy hoop of iron, which had been at one time round the rudder head, bowsprit end, or mast head of a vessel, as he said it had been taken off a large stick. I did not buy it from him, as he was in disgrace for having disappointed me about the oil. About 1 P.M. on the same day a number of the natives paid us a visit, among whom were Ec-vu-chi, I-vit-chuk, and Ou-too-ouniak, three of the most decent and best behaved of the party. They brought us a quantity of venison, of which they had still a large stock, and some of which they were now willing to dispose of, as they found that they had more than was requisite for their own consumption.

They had frequently seen Ooblooria, Ikmallik, and some of the other Esquimaux mentioned by Sir John Ross, and I also further learnt that the man with the wooden leg, named Tulluahiu, was dead, but how long since I could not discover.

The greater part of the men had been employed for the last fourteen days digging away the snow from the boat to relieve her from the pressure, as she was covered up to the depth of more than twelve feet. This was no easy task; however, we managed it in the following manner. Having cut a narrow opening through the snow down to the boat, we erected a tackle over it and hoisted up the loose snow, as it was removed with spades and axes. After excavating a space the full length of the boat, and clearing the snow out of it, the bow and stern were alternately raised, and the blocks of snow which were chopped from the top pushed underneath to prevent its sinking down again. In this way the men could work without exposure, and when the weather was stormy the hole was covered with a sail, so that the snow-drift could not interfere with our labours. We had yesterday got her close to the top of the snow roof, and to-day the weather being fine she was hauled out and found to be uninjured, except a small split in one of her thwarts caused by the great weight. She was now placed in a situation where there was no danger of her being again drifted over.

The Esquimaux left us on the 17th, having behaved themselves in the most exemplary manner. One of Akkeeoulik's wives (quite a young dame with a most interesting squint) took this opportunity of leaving her husband and putting herself under the care of her father Outoo-ouniak, the alleged cause of her dissatisfaction being that she did not get enough to eat. The disconsolate man followed the party for some distance in hopes of persuading the runaway to return, but without success.

Our fuel getting rather scarce, some of the men were sent to dig among the snow for moss and heather, and they usually got as much in a day as would cook one meal, but as the spring advanced, and the snow began to disappear, two men could procure as much as we required. When the men were taking a walk after divine service on the 21st, they saw the traces of five deer going northward.

On the 22d Turner commenced making two sledges for our spring journeys. They were to be from 6 to 7 feet long, 17 inches broad, and 7 inches high. The only wood we had for this purpose was the battens with which the inside of our boats was lined, it being necessary to nail three of them together to form runners of the required height. A wolf was shot by Ouligbuck during the night within 10 yards of the door of the house, and six or eight more were seen at no great distance off in the morning.

23rd.—When taking my usual exercise, I came upon a white owl feasting on a hare which it had killed after a severe struggle, to judge by the marks on the snow. Half of it was already eaten. Another wolf was shot on the 25th at a set gun, but there was nothing of him to be found in the morning except a little hair and blood, all the rest having been eaten or carried off by his companions. Some more deer tracks were seen going northward. On the 26th the height of Beacon Hill was found to be 238 feet above the level of the sea at half-flood.

Next day Nibitabo saw thirty deer and ten partridges, but only shot two of the latter. The former were in the middle of a large plain, and took good care to keep out of gun-shot, much to the annoyance of our deer-hunter, who is one of the keenest sportsmen I have ever met with.

There were two wolves wounded by Ouligbuck's gun last night, one of which he caught before breakfast. I went with him after the other in the forenoon, and got sight of him about three miles from the house. Although his shoulder was fractured, he gave us a long race before we ran him down, but at last we saw that he had begun to eat snow, a sure sign that he was getting fagged. When I came up with him, so tired was he that I was obliged to drive him on with the butt of my gun in order to get him nearer home before knocking him on the head. At last we were unable to make him move on by any means we could employ.

Ferocity and cowardice, often if not always, go together. How different was the behaviour of this savage brute from that of the usually timid deer under similar circumstances. The wolf crouched down and would not even look at us, pull him about and use him as we might, whereas I never saw a deer that did not attempt to defend itself when brought to bay, however severely wounded it might be.

On the 1st March one of our sledges that had been finished was tried, and found to answer well.

The deer were now steadily migrating northward, some being seen every day, but there were none killed until the 11th, when one was shot by Nibitabo; it proved to be a doe with young, the foetus being about the size of a rabbit. The sun had so much power that my blankets by being exposed to the air got completely dried, being the first time that they had been free from ice for three months.

Shortly after divine service on the 14th, Akkeeoulik, who had gone some days before to his father-in-law's to endeavour to reclaim his better half, returned with his lost treasure, one of the most lazy and dirty of the whole party, and a most arrant thief to boot. Two deer were shot on the 15th, and two more on the 18th. Deer-hunting had become very different from what it was in the autumn.

The greater part of the hollows which favoured our approach in the latter season were now filled up with snow, which, from wasting away underneath, made so much noise under foot that in calm weather it was almost impossible to get within shot. The deer were besides continually moving about in the most zig-zag directions, and were so much startled at the report of a gun that it was evident they had been a good deal hunted during the winter.

On the 20th Nibitabo was affected with inflammation of the eyes, which was relieved by dropping laudanum into them. On the 26th we made a new water hole on the lake, when the ice was found to be 6 feet 10 inches thick. I measured the dimensions of an Esquimaux canoe, and found them as follows:—length 21 feet, breadth amid-ships 19 inches, and depth where the person sits 9½ inches. The timbers are one-half or five-eighths of an inch square, and placed three inches apart near the centre of the canoe, but gradually increased to five inches at each end. The cross-bars are three-quarters of an inch thick and a foot from each other; these were morticed into gunwales 2½ inches broad by half-an-inch thick, the whole being covered with seal skin in the usual manner. Altogether it was much more neatly finished and lighter than any I had seen in Hudson's Straits; but the natives here have not attained the same dexterity in managing them, as they cannot turn their canoes without assistance after being capsized.

On the 31st Ouligbuck, who had been absent all night, came home at 1 P.M. very faint from the effects of a severe wound he had received on the arm by falling on a large dagger which he usually carried. On cutting off his clothes I found that the dagger had passed completely through the right arm a couple of inches above the elbow joint.

In the evening Shimakuk, who is a conjuror, came in, and as Ouligbuck wished to try the effect of his charms on the injured part, I of course had no objections. The whole process consisted in putting some questions (the purport of which I could not learn) to the patient in a very loud voice, then muttering something in a very low tone, and stopping occasionally to give two or three puffs of the breath on the wounded arm. During these proceedings the men could with difficulty keep their gravity; nor could I blame them, for the scene was irresistibly ludicrous.

I observed that one of the conjuror's dogs was lame, or rather very weak in the legs, and on asking him the cause, he said that it arose from having eaten trout livers when young.

The latter part of the month of March was spent, by the majority of the party, in making preparations for our journey, over the ice and snow, to the northward, it having been my intention to set out on the 1st April; but the accident to Ouligbuck prevented this, as I did not wish to leave him until I saw that his wound was in a fair way of healing. Ivitchuk, our intended companion, had not yet made his appearance.

On the 3rd April the thermometer rose above zero, for the first time since the 12th December.

As the aurora was seldom noticed after this date, I may here make a few remarks on this subject. It was often visible during the winter, and usually made its appearance first to the southward in the form of a faint yellow or straw-coloured arch, which gradually rose up towards the zenith. During our stay at Fort Hope I never witnessed a finer display of this strange phenomenon than I had done at York Factory, nor did it on any occasion affect the horizontal needle as I had seen it do during the previous winter there.

The Esquimaux, like the Indians, assert that the aurora produces a distinctly audible sound, and the generality of Orkneymen and Zetlanders maintain the same opinion, although for my own part I cannot say that I ever heard any sound from it. A fine display, particularly if the movements are rapid, is very often succeeded by stormy or snowy weather, but I have never been able to trace any coincidence between the direction of its motions, and that of the wind.

CHAPTER SIX

Set out for the north—Equipment of the party—Snow-blindness—Musk-ox—Mode of killing it—Reach the coast near Point Hargrave—Ice rough along shore—Pass Cape Lady Pelly—Unfavourable weather—Slow progress—Put on short allowance—River Ki-ting-nu-yak—Pemmican placed *en cache*—Cape Weynton—Colville Bay—High hill—Dogs giving way—Work increased—Snow-house-building—Point Beaufort—Point Siveright—Keith Bay—Cape Barclay—Another *cache*—Leave the coast and proceed across the land—River A-ma-took—Dogs knocked up—Lake Ballenden—Harrison islands—Party left to procure provisions—Proceed with two of the men—Cape Berens—Relative effects of an eastern and western aspect—Halkett Inlet—Reach Lord Mayor's Bay—Take formal possession of the country—Commence our return to winter quarters—Friendly interview with the natives—Obtain supplies of provisions from them—View of Pelly Bay—Trace the shore to the eastward—Travel by night—Explore the coast of Simpson's Peninsula—Arrive at Fort Hope—Occurrences during the absence of the exploring party—Character of the Esquimaux Ivitchuk.

Everything having been for some days in readiness for our contemplated journey, I only awaited the arrival of our Esquimaux ally Ivitchuk. He made his appearance on the 4th April in company with his wife, his father and brother, and their wives. I could have well dispensed with the presence of the party, excepting the man who was to go with us, as there were many things to be attended to. It is strange that throughout the winter, with one or two exceptions, the visits of these people have happened on Sundays. Our intended travelling companion having received a coat from one, inexpressibles from another, leggings from a third, &c., was soon completely dressed "à la voyageur," not certainly to the improvement of the outer man, but much to his own satisfaction.

Ouligbuck's arm being now in a fair way of recovery, there was no cause of detention.

The party, consisting, besides myself, of George Flett, John Corrigal, William Adamson, Ouligbuck's son, and Ivitchuk, started early on the morning of the 5th. We were accompanied by two sledges, each drawn by four dogs, on which our luggage and provisions were stowed. Our stores consisted of three bags of pemmican, seventy reindeer tongues, one half-hundred weight of flour, some tea, chocolate, and sugar, and a little alcohol and oil for fuel. At first the weather was far from favourable for travelling, as there was a gale of wind with snow, but about 8 A.M. the sky cleared up, and the day became as fine as could have been wished. The sun shone forth with great brightness, surrounded by a halo of the most brilliant colours, with four parhelia that rivalled the sun himself. Our route was the same as that followed in the boat last autumn, but although the snow was hard-packed and not rough, our sledges were too heavy to allow us to travel quickly. Numerous bands of deer crossed our path, and enlivened the scene at the same time that they kept up the spirit of our dogs. Our latitude at noon, by an observation of the sun, was 66° 42' N., variation of the compass 64° W. Between 7 and 8 P.M., both dogs and men being somewhat fatigued with their day's work, we stopped on the east side of Christie Lake to build our snow hut, which our Esquimaux friend was so long in completing on account of the bad state of the snow for building, that it was 11 o'clock before we got into our blankets. The situation of our encampment was in latitude 66° 49' 30" N., longitude 87° 20' W.

6th.—We passed a comfortable night, and it was 6 o'clock in the morning before we were again on the march; three hours more brought us to the northern extremity of the lake, where we had left a bag of flour "en cache" the previous autumn. Two men who had accompanied us, for the purpose of taking the flour back to our winter quarters, returned from this place.

A little before noon we arrived at the snow hut of the two Esquimaux, Shimakuk and Kei-ik-too-oo, who, with their families, had been staying some time here angling trout. I had agreed with those people that they should build a large snow house for our accommodation, having expected to reach them at the end of our first day's journey. In this we were disappointed; but, as the contracting party had prepared a fine roomy dwelling for us, they received the stipulated price—a clasp knife. At noon, when still on the lake, the latitude 66° 58' 16" N. was observed.

Kei-ik-too-oo having come with us for a short distance, I proposed that he should get his sledge and dogs and accompany us for two days; this, for a dagger as a consideration, he gladly agreed to do, and immediately went off at a great rate to bring up his team. Being quite light he soon overtook us, and was not long in getting a heavy load on. I soon saw the advantage of his iced runners over the iron ones, and determined to have ours done in the same way on the first opportunity; on this account we stopped sooner than we would otherwise have done, having travelled sixteen geographical miles. We found a number of old Esquimaux houses, one of which we prepared for our use by clearing out the snow that had drifted into it. Whilst the two Esquimaux were icing the sledges, the remainder of the men were cooking and preparing our bed; the latter being a very simple process, merely requiring the snow to be well smoothed, and one or two hairy deer-skins laid over it to prevent the heat of the body from thawing the snow. The weather was fair all day, and except in the morning when the thermometer was -16°, it was rather warm for walking. After we got into our lodgings a strong breeze sprung up with thick drift. Some of the party were slightly affected with snow-blindness.

7th.—The weather was gloomy and dark this morning, with the thermometer at +5° when we started at half-past 3. Our sledges ran much easier since they had received a coating of ice on their runners, although they were not yet equal to Kei-ik-too-oo's. We followed the same route as that taken by the boat last autumn until 9 o'clock, when being two miles from the sea we struck across land towards Point Hargrave; at noon we were in latitude 67° 16' 51" N., variation of the compass 74° 30' W. We found the snow much softer than it was on the lakes and river, and our progress was consequently much slower than in the first part of the day.

At 2 P.M. we arrived at a small lake, about four miles from Point Hargrave. As this was the only fresh-water lake we were likely to meet with for some time, I determined to stop for the purpose of renewing the icing on the sledges, which had been a good deal broken by the irregularities of the road. Notwithstanding that we had gone only eighteen miles our dogs were very tired, and I began to fear that they would not hold out so well as was expected. Our Esquimaux friend was to leave us the next day, and as his sledge was light he expected to reach his house the same day. This is a favourite resort of the musk-ox as soon as the snow disappears. The mode of killing these animals is the same as that described by Sir J.C. Ross as practised in Boothia Felix by the Esquimaux: being brought to bay with dogs, they are either shot with arrows or speared.

When we resumed our journey at 5 o'clock next morning, there was a strong breeze right ahead with thick drift, the temperature being +6°. A walk of three miles brought us to the coast about a mile from Point Hargrave. There was a great deal of rough ice along the shore, which gave both men and dogs much hard work to drag the sledges over. It had now begun to snow, and the drift was so thick that we could not follow the smoothest route; we consequently advanced but slowly, taking four hours to gain five and a half miles, which brought us to Cape Lady Pelly.

Since leaving Fort Hope, I had measured every foot of the ground we had passed over with a line, but now the increased difficulty of the route made it requisite that all hands should be employed in dragging the sledges. One of our best dogs became quite useless, and although unharnessed would not walk, so that rather than lose the poor animal, we dragged him on the snow several miles before reaching our intended encampment.

After passing Cape Lady Pelly the coast turns rather more to the westward. The weather continued very unfavourable all day, there being much snow-drift; we however advanced seven miles farther, and at 4 P.M. built our night's lodgings on the ice, a few hundred yards from the shore. In an hour and a half we were comfortably housed. Finding that our day's journeys were much shorter than I had anticipated, our allowance of food for supper was somewhat reduced. The thermometer in the evening stood at +11°. Our snow hut was situated in latitude 67° 35' N., longitude 87° 51' W. both by account.

After a sound night's rest we resumed our journey at 5 in the morning of the 9th. There was some snow falling, but the wind had decreased, and the temperature of the air was +2°. Our course was N.W. by W. for three miles, when we came to a low point formed of shingle and mud, with some rocky rising grounds a few miles inland. This point received the name of Swanston, after a friend. A short time before noon the sky cleared, and very satisfactory observations for latitude and variation of the compass were obtained, the former being 67° 40' 53" N., the latter 71° 30' W. The dog that had been unharnessed the day before had become still weaker, and as I did not wish to leave him to the mercy of the wolves, he was shot. We offered some of his flesh to the other dogs, but there was only one of them that would eat it.

Having walked fourteen miles, we arrived at a small river 70 yards wide, and, although it was only half-past three, we commenced building our snow house. We here found a number of stones which allowed us to place "en cache" half a bag of pemmican, some flour, shoes, &c., for our homeward journey. The river, which is called Ki-ting-nu-yak, was frozen to the bottom, but in summer it is a favourable fishing station, both salmon and a small species of the white fish being found. I did not see any of the latter, but from the description given by the Esquimaux I have no doubt that they frequent this part of the coast.

The evening was beautifully clear, and the thermometer fell to -16°.

10th.—There was a thick haze this morning with light variable airs of wind; temperature 6° below zero. By striking straight out from land for a mile or two, we got upon somewhat smoother ice, and consequently made more progress. We passed a number of hills, not of any great elevation however, and at noon we were opposite one named Wiachat, fully 500 feet high, and some miles from the coast. Here the latitude 67° 53' 24" was observed, and the coast turned off to the westward, forming a point which was named Cape Weynton. We now commenced crossing a bay 5 or 6 miles deep, and apparently 12 wide, which received the name of Colville, in honour of the Deputy Governor of the Hudson's Bay Company. A mouse or lemming crossed our path, and the dogs, although they appeared to be scarcely able to put one foot before another, set off at full speed in chase, and before any one could interfere to save it, the poor little animal was quivering in the jaws of the foremost.

Being unable to reach the north side of Colville Bay, at 4 P.M. we took up our quarters on the ice in our usual snug lodgings, in latitude 68° 2' N., longitude 88° 21' W. A high hill bearing west of us, and distant eight miles, called Oo-me-we-yak by the natives, was named after the late John George McTavish, Esq., Chief Factor. Several of our dogs had become very weak—so much so that during the latter part of the day's journey they did little or nothing, thus giving us all much additional work. They also required much more food to keep them in good condition, than the dogs generally used in the fur countries. We only walked sixteen miles this day; and I may here remark that all the distances mentioned in this journal are given in geographical miles.

Our usual mode of preparing lodgings for the night was as follows:—As soon as we had selected a spot for our snow house, our Esquimaux, assisted by one or more of the men, commenced cutting out blocks of snow. When a sufficient number of these had been raised, the builder commenced his work, his assistants supplying him with the material. A good roomy dwelling was thus raised in an hour, if the snow was in a good state for building. Whilst our principal mason was thus occupied, another of the party was busy erecting a kitchen, which, although our cooking was none of the most delicate or extensive, was still a necessary addition to our establishment, had it been only to thaw snow. As soon as the snow hut was completed, our sledges were unloaded, and every thing eatable (including parchment skin and moose skin shoes, which had now become favourite articles with the dogs) taken inside. Our bed was next made, and by the time the snow was thawed or the water boiled, as the case might be, we were all ready for supper. When we used alcohol for fuel (as we usually did in stormy weather) no kitchen was required.

On the following morning we started about the usual hour, and directing our course nearly north, a walk of five miles brought us to the opposite side of Colville Bay, which terminated in a long point covered with boulders of granite and debris of limestone, and having a number of stone marks set up on it. To this point the name of Beaufort was given, in honour of the gallant officer who, with so much advantage to his country and to nautical science, presides over the hydrographical department of the Admiralty.

Five miles farther we reached another low point called by the Esquimaux E-to-uke, but renamed by me Point Siveright. The coast, now turning slightly to the westward of north, continued in nearly a straight line during the rest of this day's march.

We were now tracing the shores of a considerable bay, as the land after taking a sudden bend to the eastward followed a south-east direction as far as visible. At 4 P.M. we stopped and built our snow hut; the day had been fine throughout, and the temperature in the evening was 16° below zero. The shores of the bay are very low, with the exception of a high bluff point bearing S.E. by E. 6½ miles (by trigonometrical measurement). The point was named Cape Barclay, in honour of the Secretary of the Hudson's Bay Company, and the bay was called after my much respected friend, George Keith, Esq., Chief-Factor.

Since passing Colville Bay the coast had become much lower and more level, giving every indication of a lime-stone country. Being anxious to save our fuel as much as possible, we filled two small kettles and a bladder with snow and took them to bed with us, for the purpose of procuring water to drink—a plan which was frequently adopted afterwards. Our dogs had now become most ravenous; although they received what was considered a fair allowance of provisions, everything that came in their way, such as shoes, leather mitts, and even a worsted belt, was eaten, much to the annoyance of the owners and to the merriment of the rest of the party. We enjoyed a cold supper of pemmican and water;—as we could afford a hot meal only once a day, we preferred taking it in the morning.

12th.—Being informed by our Esquimaux companion that, by crossing over land in a north-west direction to a large bay which he had formerly visited, we should shorten our distance considerably, I determined on adopting the plan proposed. Our kettles of snow were found rather cool companions, but there was a little water formed. The bladder having been either leaky, or not properly tied, gave me and my next neighbour a partial cold bath. The morning was delightful, being clear and calm, with a temperature of -22°. We started at half-past 5, and after having walked a short distance came to some loose pieces of granite and limestone, which afforded an opportunity, not to be lost, for making a deposit of provisions for our return journey.

After tracing the shores of the bay for three miles and a half further north to latitude 68° 18' N., longitude by account 88° 26' W., we left the coast and proceeded over land in a north-north-west direction. Walking became more difficult, and the snow was too soft to support the sledges, the ice on the runners of which was now entirely worn off. A mile's walk brought us to a small river with high mud banks, and frozen to the bottom: it is named A-ma-took by the natives, and takes its rise from a lake of the same name about a day's journey west of us. We next passed between two elevations covered with limestone. I ascended that on the right-hand or to the east of us, it being the highest and having two columns of limestone, the one fourteen, the other nine feet high, at its western extremity. There were many places here denuded of snow, showing that the sun had already acquired great power. At noon we were in latitude 68° 22' 19" N., variation of the compass 79° 35' W. An hour after, we reached a small lake, where we halted on account of our dogs being quite knocked up, although we had only advanced twelve miles; I therefore ordered a hut to be built that we might afford the dogs time to recruit, and also have the sledge-runners put in order. We found the ice on the lake 4 feet 8 inches thick, but we were disappointed to find that there were no fish to be caught. We here enjoyed water ad libitum, a luxury that had been rather sparingly dealt out for the last few days. Ivitchuk drank as much as would have satisfied an ox. The thermometer in the evening was 9° below zero. A few tracks of foxes were here seen, but no signs of deer or musk-oxen. This part of the country appeared miserably barren in every respect.

On the morning of the 13th we commenced our march at 2.30 A.M.; the weather was fine with light airs from the north-west: thermometer -15°. At 5 o'clock we passed a small lake about a mile and a half long, and an hour afterwards reached another of considerable size. Tung-a-lik, as the lake is called by the Esquimaux, is 7 miles long due north and south, and varying in breadth from a mile to a mile and a half. Near its centre was a curious-looking island, about 7 feet high and 200 yards in extent, covered with granite boulders and limestone. Its form is as nearly as possible that of a semicircle, the concavity being towards the south. To this lake I gave the name of Ballenden, after a much valued friend. When near the north end of Ballenden Lake (over which we had travelled rapidly, the snow being both hard and smooth), we turned more to the west. At noon we arrived at a lake which was to be our resting place for the night, as, although small, it was said to contain both trout and salmon, but, after cutting through five feet of ice, we did not succeed in catching any, although we tempted them with a bait from a buffalo hide. In the afternoon the weather became very gloomy; a strong breeze sprang up accompanied by a thick haze, and the thermometer rose to -11°. By meridian observation our latitude was 68° 36' 58" N., variation of the compass 78° W., longitude by account 88° 49' W.

14th.—This morning was so stormy, with thick drift and snow, that we could not start so early as usual; it however became more moderate at 5 o'clock, and we were able to continue our route, although the guide seemed much puzzled to keep in the proper direction, there being nothing to serve as marks in this wilderness of snow.

In the afternoon the weather again became worse, and the temperature fell to -12°, which with a strong head wind made it sufficiently cold.

I felt it probably more than the others, as I had to stop often to take bearings, and in consequence was once or twice nearly losing the party altogether. We trudged on manfully until 5 P.M., when it cleared up for two or three minutes, and we obtained a distant glimpse of some high islands in the bay for which we were bound, called Ak-koo-lee-gu-wiak by the natives. At half-past 5 we commenced building our snow house. This was far from pleasant work, as the wind was piercingly cold, and the fine particles of snow drift penetrated our clothes everywhere; we, however, enjoyed ourselves the more when we got under shelter and took our supper of the staple commodities, pemmican and water. Latitude 68° 51' N., longitude 89° 16' W.

15th.—It blew a complete storm all night, but we were as snug and comfortable, in our snow hive, as if we had been lodged in the best house in England. At 5.30 the wind moderated to a gale, but the drift was still so thick that it was impossible to see any distance before us, particularly when looking to windward, and that unfortunately was the direction in which we had to go. The temperature was 21° below zero,—a temperature which, as all Arctic travellers know, feels much colder, when there is a breeze of wind, than one of -60° or -70° when the weather is calm. But there was the prospect of both food and fuel before us, for seals were said to abound in the bay and heather on the islands of Akkooleeguwiak. Such temptations were not to be resisted; so we muffled ourselves well up and set out. It was one of the worst days I had ever travelled in, and I could not take the bearings of our route more than once or twice.

To make matters worse, one of our dogs, a fine lively little creature, that was a great favourite with us all, became unable to walk unharnessed, and the men having enough to do with the sledges, I dragged, carried, and coaxed it on for a few miles; but finding that some parts of my face were freezing, and that my companions were so far ahead as to be out of sight, I was reluctantly compelled to leave the poor animal to its fate.

After a most devious course of nearly twelve miles, we came to the shores of the bay. The banks were of mud and shingle, about sixty or seventy feet high, and so steep that it was some time before we could find a place by which to get down to the ice. We directed our steps among much rough ice towards the highest of the group of islands named Coga-ur-ga-wiak, apparently six miles distant, and encamped near its western end in a little well-sheltered bay. All the party, even the Esquimaux, had got severely frost-bitten in the face, but as it was not much more than skin deep, this gave us little concern. When our house was nearly built, a search was commenced among the snow for heather, and we were so fortunate as to procure enough in an hour and a-half to cook us some pemmican and flour, in the form of a kind of soup or pottage.

We were all very glad to get into our blankets as soon as possible. The weather became somewhat finer in the evening, but it drifted as much as ever. The thermometer was -16°. Our latitude was 68° 53' 44" N., longitude 89° 55' 30" west. Notwithstanding that I carried my watch next my skin the cold stopped it, and I could not tell exactly the time of our arrival at the island, but I believed it was near 2 P.M.

On the 16th, a gale of wind from N.W. with thick drift, and the thermometer at -20°, would have prevented our travelling had I intended it; but as I purposed leaving some of the men and all the dogs here to recruit, I wished to find out the Esquimaux (who we knew were in the neighbourhood, as the recent foot tracks of two had been seen on the shore the day before), and obtain from them some seals' flesh and blubber for our use. Flett, Ivitchuk, and the interpreter were sent on this mission, but they returned in the evening unsuccessful. The drift was so thick in the bay that they could not see to any distance. In the meantime Corrigal and Adamson had been collecting fuel, and I being under the lee of the island obtained observations for latitude and variation of the compass, the former being 68° 53' 44" north as above, the latter 87° 40' west.

I prepared for an early start the next morning in company with Flett and Corrigal, for the purpose, if possible, of reaching Sir John Ross's most southerly discoveries, which could not now be distant more than two days' journey.

The party that were to be left behind had orders to kill seals, (for which purpose Ivitchuk was furnished with a spear,) to trade provisions from the Esquimaux if they saw any, and, above all, to use as little of our present stock as possible. All that we could afford to take with us was four days' scanty allowance. I had for the last week carried my instruments, books, &c., in all about thirty-five pounds weight; and I now intended to do the same.

The morning of the 17th was stormy and cold (-22°), and we did not start until near 6 o'clock; when we had got well clear of the S.W. end of the island, we found the ice smooth, and the snow on it hard-packed. As the men had but a light load we travelled fast, our course being nearly N.W. towards the farthest visible land in that direction. A brisk walk of seventeen miles brought us, an hour before noon, to the shore near a high point formed of dark gray granite, which I named Cape Berens, after one of the Directors of the Company. It is situated in latitude 69° 4' 12" N. by observation, and longitude 90° 35' 48" W. by account. The shore, which was steep and rocky, ran nearly in a straight line, and in the same direction that we had been already travelling. At 3 P.M. we came to two narrow points in a small bay, between which we built our snow-house. To these points I gave the name of "the Twins." Their latitude is 69° 13' 14" N., longitude 90° 55' 30" W.

There being one or two hills at a short distance from us, I ascended one of them to look for fuel, and to gain a view of our future route. I obtained neither of these objects, but fell in with some lead ore, specimens of which were brought away.

On arriving at the snow-hut I found it nearly completed, but so small that there was little prospect of a comfortable night's rest. Having but a very small quantity of alcohol for fuel, our supper was a cold one. Thermometer in the evening 19° below zero. Flett (one of Dease and Simpson's best men) showed symptoms of fatigue, at which I was much surprised, as, from what I had heard of him, I fancied he would have tired out any of the party.

18th.—My anticipations of passing an uncomfortable night were fully realized. It might be thought that, as our whole bedding consisted of one blanket, and a hairy deer-skin to put between us and the snow, there was reason enough for my not sleeping soundly; but this was not the case, as I often passed nights both before and after this with as little covering, but never found myself cold. We started at 3 A.M. The morning was fine but hazy, with a light air of wind from N.W. Thermometer -3°. The walking was still fair; and I may here remark, that wherever the land had an eastern exposure the ice was smooth, there being little or none of the former year forced up along the shore; whenever the coast was exposed to the west, the contrary was the case.

Our course was nearly that of the previous day, but a little more to the westward. After walking twelve miles we came to what proved to be the head of a deep inlet, the western shore of which we had been tracing, and which I named after John Halkett, Esq., one of the Directors of the Hudson's Bay Company, whose son (Lieut. P.A. Halkett, R.N.,) is the ingenious inventor of the portable air-boat, which ought to be the travelling companion of every explorer. Two reindeer were seen here.

As there could be no doubt that, if my longitude was correct, I must now be near the Lord Mayor's Bay of Sir John Ross, I decided on striking across land, as nearly north as possible, instead of following the coast. The men having had a short time to rest, we commenced a tiresome march over land, the snow being in some places both deep and soft. We crossed three small lakes, and at noon, when near the middle of another about four miles long, an excellent meridian observation of the sun gave latitude 69° 26' 1" N. When we had walked three miles more we came to another small lake; and here, as there was yet no appearance of the sea, I ordered my men to prepare our lodgings, whilst I went on alone to endeavour to discover the coast.

A walk of twenty minutes brought me to an inlet not more than a quarter of a mile wide. This I traced to the westward for upwards of a league, when my course was again obstructed by land. There were some high rocks near at hand which I ascended, and from the summit I thought I could distinguish rough ice in the desired direction. With renewed hopes I slid down a declivity, plunging among snow, scrambling over rocks, and through rough ice until I gained more level ground. I then directed my steps to some rising ground which I found to be close to the seashore. From the spot on which I now stood, as far as the eye could reach to the north-westward, lay a large extent of ice-covered sea, studded with innumerable islands. Lord Mayor's Bay was before me, and the islands were those named by Sir John Ross the Sons of the Clergy of the Church of Scotland.

The isthmus which connects the land to the north with the continent is only one mile broad, and even in this short space there are three small ponds. From the great number of stone marks set up (the only ones that I saw on this part of the coast), I am led to infer that this is a deer-pass in the autumn, and consequently a favourite resort of the natives. Its latitude is 69° 31' N., longitude by account 91° 29' 30" W. This latter differs only a mile or two from that of the same place as laid down by Sir James C. Ross, with whose name I distinguished the isthmus, calling the land to the northward Sir John Ross's Peninsula. After going down to the ice in Hardy Bay, and offering with a humble and grateful heart thanks to Him who had thus brought our journey so far to a successful termination, I began to retrace my steps towards my companions.

At a late hour I reached our snow hut, an excellent roomy one, in which we could lie in any position; no trifling comfort after a walk of more than forty miles over a rough road.

It was 7 o'clock the following morning before we started. The weather was pleasant, and the thermometer 12° below zero. Having taken possession of our discoveries with the usual formalities, we traced the inlet eastward, the shores of which were steep and rugged, in some places precipitous. When we had walked four miles the land on our left turned up to the northward, leaving an opening in that direction more than two miles wide, bounded on the south-east by one or more islands. This inlet I named after that celebrated navigator and discoverer Sir John Franklin, whose protracted absence in the Arctic Sea is at present exciting so much interest and anxiety throughout England. The most distant visible point was called Cape Thomas after a relative. The land on our right still trended to the east for two miles, and then turned to the south. After walking seven miles in this last direction, and passing two small bays and as many points, we stopped for the night. Here we were fairly puzzled about the proper route, there being so many inlets and small bays that it was impossible to tell which was the one we ought to follow. The day had become very warm, the thermometer rising as high as +26° in the sun, and as we were now travelling south, we found the reflection from the snow much more painful to the eyes than when proceeding north. The latitude of our snow house was 69° 22' N., longitude 91° 3' W., both by account. The thermometer -19° in the evening; cold water and pemmican for supper, and kettles of snow for bedfellows.

The morning of the 20th was cold, but calm; thermometer -24°. We commenced our day's march at 2h. 30m. A.M., and in twenty minutes arrived at the head of the inlet where I hoped to find a passage. Seeing that it would be madness to trace all the indentations of this most irregular coast, (for had a couple of days' stormy weather ensued we should all have run the risk of starving,) I struck over land towards our snow hut of the 17th.

This was the most fatiguing and at the same time the most ludicrous march we had experienced. As our route lay across several ranges of hills, we had no sooner climbed up one side than we had to slide down the other. To descend was not always an easy matter, as there were often large stones in the way, past which we required to steer with great care, or if a collision was unavoidable, to manage so as not to injure ourselves. Corrigal appeared to be an old hand at this sort of work, and I had had some practice, but poor Flett, who had begun to suffer much from inflammation of the eyes, got many queer falls, and was once or twice placed in such situations with his head down hill, his heels up, and the strap of his bundle round his neck, that it would have been impossible for him to get up by his own unaided exertions.

After crossing a number of small lakes, we arrived at the steep shores of Halkett Inlet about 11 o'clock, having been eight hours in walking as many miles. We crossed the inlet, and as it had now begun to blow a fresh breeze we stopped at a small bay, well sheltered, to take some rest, and obtain a meridian observation of the sun. The latitude was

69° 16' 44" N., variation of the compass 76° 45' W. We were so fortunate as to find here some heather by scraping away the snow, and we enjoyed the luxury of a cup of chocolate, which refreshed us very much.

We now resumed our march, and the walking being good and the day fine we made rapid progress, although somewhat detained by the lameness and blindness of Flett, who stumbled at every inequality of the ground, and received some severe falls. After advancing two miles we came opposite to a clear opening to the north-eastward, in which nothing but rough ice could be seen. This was evidently the termination of the continent in this direction. At 4 P.M. we arrived at our snow hut in the small bay between the Twins. It was not my intention to remain here all night, but the lameness of our companion prevented us from continuing our journey. Whilst I went to search for fuel, Corrigal enlarged our snow house. I found a little fuel, with which we contrived to thaw as much snow as gave each of us nearly half-a-pint of water. The remainder of our provisions, amounting to a few ounces of pemmican each, was fairly divided, and having eaten part of this we betook ourselves to rest.

21st.—Having passed a far from pleasant night, and used the last of our alcohol to procure some water as a diluent for our not very plentiful breakfast, we started at a little before 2 A.M. There was a strong breeze from N.W. with thick drift occasionally, and a temperature of -20°, but the wind being on our backs it was rather an advantage than otherwise. We directed our course straight for the island on which we had left the rest of the party, and which could be seen at intervals when the snow drift cleared away.

Flett being still very lame, I desired Corrigal to remain in company with him, whilst I went on alone to order some provisions to be prepared by the time they came to the snow house. The ice being smooth, and the snow on its surface hard, I made rapid progress until within about five miles of our temporary home. Here I observed some strange looking figures on the ice, which the thickness of the weather prevented me from seeing distinctly. On a nearer approach I found that what had puzzled me was a number of Esquimaux spears, lances, &c. stuck on a heap of snow, and immediately afterwards four Esquimaux came from behind a mound of ice, holding up their hands to show that they were unarmed. The natives of this part of the coast bear a very bad character, and are much feared by their countrymen of Repulse Bay. I therefore was not quite sure what sort of reception I might meet with, as my men were not in sight and I was quite unarmed. But to anticipate evil is often the most likely way to cause it, so I went directly up, and saluted them with their usual term for peace (teyma), shaking hands with all after the fashion of our own country. They all shouted out Manig Tomig, which are the words mentioned by Sir John Ross as the form of salutation employed by the natives of Boothia Felix. A very animated conversation soon ensued, in which I bore but a very small share; but as I appeared to be a good listener, and put in a negative or affirmative every now and then when there appeared to be a necessity for saying something, we got on very well together.

We were soon joined by an old woman who took upon herself the office of mistress of the ceremonies, and commenced with great volubility to give me the names of the men, which were as follows:— A-li-ne-a-yuk, Kag-vik, Tag-na-koo and Nu-li-a-yuk; the first being old, the second middle aged, and the two last young men of about twenty-five. They were all married, and were much more forward in their manners and dirty in their persons and dress, than our friends of Repulse Bay. They were very anxious for me to enter their huts, but this I thought it prudent to decline, and after much persuasion and promises of knives, needles, beads, &c. I prevailed on them to follow me to our snow house.

A little more than an hour brought me to our encampment, where I found Adamson quite well but all alone, Ivitchuk and the boy being out looking for seals. They had not met with any Esquimaux, and no animals of any kind had been killed, Ivitchuk standing so much in awe of his countrymen that he was afraid to stay out seal-hunting during the night, which is the only time that these animals are to be caught at that season of the year. I found that much more of our stock of provisions had been used than there was any occasion for—in fact, the appearance of the men shewed that they had been on full allowance.

About an hour after my arrival, Corrigal and Flett made their appearance, accompanied by the four Esquimaux that I had seen and a boy. A few trifling presents were made them, and they promised to return on the following day with oil, blubber, &c. to barter with us. It blew a gale all the evening, with the thermometer 21° below zero.

The morning of the 22d was fine with a temperature of -20°, but during the day it blew hard with drift. Our party kept in bed rather longer than usual, and we were visited by the Esquimaux before we had got up. They brought a quantity of seal's flesh, blood, and blubber, which I was about to purchase from them when the thermometer was reported as missing. I immediately shut the box containing the valuables, and intimated that they should receive nothing unless the thermometer was given up. After about ten minutes' delay one of the women brought in the lost article, saying, that the dogs had pulled it down and carried it off,—a very probable story certainly; but having obtained what I wanted I cared little who might be the thief.

A brisk traffic was soon commenced for oil, seals, blubber, flesh and blood, for which knives, files, beads, needles, &c. were given. We also obtained half a dozen dried salmon and a small piece of dried musk-ox flesh, both very old and mouldy. These Esquimaux were found to be much more difficult to deal with than our friends of Repulse Bay, being very forward and much addicted to stealing. They had undoubtedly had communication with the natives of Boothia Felix, as there were many of their weapons, and parts of their sledges formed of oak. I also observed some small pieces of mahogany among them. One of the strangers proved to be an uncle of Ivitchuk.

It continued to blow hard in the evening with a temperature of -15°. Preparations were made for examining the shores of the bay in which, by Esquimaux report, we now were.

23rd.—This was another stormy and cold day until the afternoon, when it became fair. We were again visited by our neighbours, who brought us a further and very acceptable supply of seals' flesh and blood, and also two fine dogs to complete our teams, one or two of those we had being still very weak.

When about to make a tour round the bay, I learnt from one of the natives that a complete view of its shores could be obtained from the summit of the island on which we were. I found also that a chart which he made of the bay agreed very closely with one drawn by the natives of Repulse Bay, who had visited the place. The evening being beautifully clear, I took with me the Esquimaux, one of the men, and the interpreter to the highest point of the island, from which I obtained a distinct view of the whole bay, except a small portion immediately under the sun. The shores were high and regular in their outline, and being, in most places, to a certain extent denuded of snow, they were much more clearly seen than could have been expected. The bay appeared to extend 16 or 18 miles slightly to the east of south, and was about 11 miles wide near its head. Its surface was studded with a number of dark-coloured rocky islands. The highest of these was the one on which we were staying, and was found by measurement to be 730 feet above the level of the sea. It was called Helen Island, whilst the group to which it belonged was named after Benjamin Harrison, Esq. one of the Directors of the Hudson's Bay Company. The Esquimaux pointed out the direction in which two rivers near the head of the bay lay. These rivers, of which I took the bearings by compass, were said to be of no great size, and frozen to the bottom in winter. The bay was honoured with the name of Sir John H. Pelly, Bart., Governor of the Company.

The morning of the 24th was as beautiful as could be desired, with the thermometer at -15°. There was a gentle air from the east, and the horizon being very clear, I again obtained a fine view of the bay.

Having abundance of blubber for dogs' meat and fuel, and as much seals' flesh and blood for ourselves as at half allowance would serve us for six or seven days, I determined to trace the shores of the land across which we had travelled on our outward journey.

For this purpose, both men and dogs being now much recruited, we started at 8h. 30m. A.M. and took a N.E. by E. course towards the eastern shore of the bay, which, having a western exposure, was much encumbered with rough ice. We had some trouble in getting over this, but found it more smooth along the shore, which trended due north. Finding that our sledges were too heavily laden, we left on the ice a quantity of our oil and blubber. Here we made a mistake in retaining the fresh fat of the seal, instead of that which had become somewhat rancid, as we found that, although the dogs ate the latter with avidity, they would scarcely taste the former. This Ivitchuk well knew, but he was too stupid to tell me of it at the time. One of our dogs that had done his work well since leaving Repulse Bay, had become so weak that he could scarcely walk. We endeavoured to coax him on, but unsuccessfully; it was therefore thought advisable to leave him where we had lightened our load, as he would have provisions for at least a fortnight, if not assisted by other animals, and before that time he would very likely be found by the Esquimaux. A meridian observation gave latitude 68° 56' 46" N., variation 78° 56' W.

As the sun had acquired too much power for travelling comfortably during the day-time, I stopped early so as to be able to continue our journey about midnight. Our snow hut was built near a small creek, in latitude 68° 58' N., longitude 89° 42' W. The coast had become low and flat, with a few fragments of lime-stone and granite boulders showing themselves occasionally above the snow. The thermometer exposed to the sun's rays rose to +37°. A little snow fell in the evening.

On the morning of the 25th there was some more snow with a temperature of -7°. We did not commence our march until some hours later than I had expected. The direction of the land continued nearly north for eight miles; it then turned off to the north-east, and continued so until we stopped at noon, in latitude by observation 69° 14' 37" N. longitude by account 89° 18' 18" W. The tracks of a large Polar bear and of some lemmings were noticed this day.

26th.—The morning was dark and cloudy when we started at 20 minutes after one. When just about to set out, we were joined by the poor dog we had left behind. He had grown into much better condition, although he was still unable to haul. I may here add that he afterwards quite recovered, and was the only one of our stock that I took to England with me.

Our course for seven miles was east, and then turned off S.E. by S. forming a cape, which was named Chapman, after one of the Directors of the Hudson's Bay Company.

We continued walking on, in nearly a straight line, for 11 miles, when our dogs became tired, and we encamped an hour before noon, in latitude by observation 69° 5' 35" N., variation 81° 50' W., longitude by account 88° 43' W. At 11 P.M. we recommenced our march, the weather being beautiful, and the temperature -8°.

27th.—The coast trended in exactly the same direction as that we had passed during the latter part of the preceding day's journey; the walking was in general good, and our dogs were every day recovering their strength. A single rock grouse (*tetrao rupestris*) was seen, but so shy that we could not get a shot at it. Many traces of foxes, and the recent foot-marks of a large white bear, were also seen. We kept a sharp out-look for the latter, with the hopes of getting a few steaks out of him, but he did not show himself. There was a high wall of broken ice all along the shore here, which may be readily accounted for by the direction of the coast, which, by contracting the bay, is exposed to the pressure of the ice coming from the northward. Fortunate it was for us that we had got some oil and seals' blubber, for there was not a bit of anything in the shape of fuel to be seen along this barren shore. The weather having become too warm, about 11 A.M., we stopped in latitude, by observation, 68° 51' N., longitude by account 88° 6' W.

The morning of the 28th was particularly fine, with a temperature of 15° below zero. For eight miles our course was the same as that of the day before, but the land now turned gradually to the southward, and finally to about a south-by-west direction. At noon the sun had become so warm, that we were compelled to encamp for the day. At three miles from where we had stopped, we passed a small bay, about 1½ mile wide, the only indentation of the coast we had seen since leaving Pelly Bay.

Our latitude by meridian observation was 68° 32' 40" N., variation of the compass 70° 55' W., and our longitude by account 88° 2' W.

29th.—We resumed our march at a little after 11 P.M. on the 28th. The weather was calm, but cloudy, with the temperature -3°. The line of coast now ran nearly south, and after a walk of five miles we came to a narrow point, extending two miles to the eastward. We then crossed a bay about 1½ mile wide, and arrived at another point of nearly the same dimensions, both formed of mud and shingle. These I named respectively after James and Robert Clouston, two intimate friends.

Four miles further brought us opposite to a small low island, half a mile from the shore, and at a short distance beyond this we came to a small bay upwards of a mile wide. A little before noon we stopped to build our snow hut. The day was now warm, the thermometer having risen as high as +55° in the sun, and +18° in the shade. One of our best dogs got lamed by putting his foot into a crack in the ice. We saw the smoke of open water at no great distance, and heard the ice making a loud noise as it was driven along with the tide. There were numerous traces of foxes, and the tracks of a band of deer, with a wolverine in pursuit, were noticed. The latitude of our position was 68° 15' N., variation 75° 52' W., and longitude by account 88° 5' 36" W.

30th.—We started at half-past nine the previous night, with clear weather and a fresh breeze from west, which, with a temperature of -8°, made our already frost-bitten faces smart severely. After a few miles' walk, we rounded a low spit of land, which had been hid from our view by the rough ice on our outward journey, and which I now named Point Anderson. Between this point and Cape Barclay, of which we now got sight, there is a narrow bay running up to the northward two or three miles.

We had a great quantity of rough ice to scramble over, which, however fatiguing, afforded some amusement, as the ridiculous positions in which we were sometimes placed gave abundant food for mirth to those who were disposed to look at every thing in the most favourable light.

About midnight the weather became very stormy, so much so indeed that we had great difficulty in keeping the proper course, which was now to the north west, for the purpose of picking up the pemmican, &c. which we had deposited on the shore of Keith Bay on the 12th. On reaching the west side of the bay at 3 A.M. I found that we were not more than a hundred yards from where our "cache" was placed, which we found quite safe. Ivitchuk and the boy having lagged behind, we removed a quantity of snow, and took possession of our old snow hut to wait for them. After staying for an hour we resumed our journey, thinking that our companions might have taken a shorter route across the bay, and this we found to be the case. It had been cold and stormy during the greater part of the night, but at 8h. 30m. A.M., when we encamped opposite Cape Beaufort, the weather had become beautiful.

The whole of the coast which we had traced during the last seven days, as far as Cape Barclay, was low and flat, with neither rock nor hill to interrupt the sameness of the landscape. It was named Simpson's Peninsula after Sir George Simpson, the able and enterprising Governor of the Hudson's Bay Company's territories, who projected and planned the expedition, and to whose zeal in the cause of discovery Arctic travellers have been so often and so much indebted.

During the remainder of our journey homewards, having followed as nearly as possible our outward route, we met with little of any interest. We reached our encampment of the 9th of April on the 1st of May, and found our "cache" of provisions quite safe. We had now an abundant stock of food, nor were we sorry to exchange the seals' flesh and blood, on which we had been subsisting for eight days past, for pemmican and flour. It is true that during that time we had supped on a few dried salmon, which were so old and mouldy that the water in which they were boiled became quite green. Such, however, is the advantage of hard work and short commons, that we enjoyed that change of food as much as if it had been one of the greatest delicacies. Both the salmon, and the water in which they were cooked, were used to the last morsel and drop, although I firmly believe that a moderately well fed dog would not have tasted either.

We now saw numerous tracks of rein-deer, all proceeding in a N.E. direction towards Melville Peninsula. Early on the morning of the 3rd of May we arrived at the small lake near Point Hargrave, on which we had encamped on the 7th of April; much of the snow had disappeared from the ground in the neighbourhood, and the marmots had already cleared out the entrances to their burrows, and recommenced their life of activity for the summer season. Not an hour now passed without our seeing deer; but they were extremely shy, and the only benefit we received from them was the life and spirit their presence infused into our dogs.

The night of the 4th was very unpleasant, there being much snow and drift, which prevented us from seeing the ridges of snow which occurred frequently on our path, and which being very hard and slippery, caused us many falls. At half-past 1 on the morning of the 5th we reached some old Esquimaux dwellings on the border of Christie Lake, about fifteen miles from Fort Hope, in one of which we took up our temporary abode. At 2 P.M. on the same day we were again on the march, and arrived at our home at 8h. 30m. P.M. all well, but so black and scarred on the face from the combined effects of oil, smoke, and frost bites, that our friends would not believe but that some serious accident from the explosion of gunpowder had happened to us. Thus successfully terminated a journey little short of 600 English miles, the longest, I believe, ever made on foot along the Arctic coast.

During our absence every thing had gone on prosperously at winter quarters. The people had been all in good health, and the wound in Ouligbuck's arm had healed up, but the limb had not yet acquired much strength. When I set out on the 5th of April there was but a very small quantity of venison in store, so that I was afraid that Folster (the man left in charge) would be forced to use pemmican, which substantial article I wished to save as much as possible for future contingencies. Fortunately the Esquimaux brought a little venison to barter, which, with an occasional deer killed by the hunters, kept the party in food; although the store at one time was so empty, that they were compelled to have a dinner of tongues, which (except in case of necessity) were to be kept for journeys. As the weather in the latter part of April became stormy, and the deer numerous, the hunters were more successful, and there was no further scarcity. Ouligbuck had, notwithstanding the wound in his arm, killed four deer, and sixteen more had been shot by Nibitabo and some others of the party; so that the meat store was well stocked when I arrived; and well that it was so, for we were as ravenous as wolves, and I believe ate more than would have been good for us had our food been anything but venison, which is so digestible that a person may eat almost any quantity without feeling any bad effects from it.

May commenced with a beautiful day, the thermometer being above zero, and continuing so throughout. This was the only day for many months past that the negative scale of the thermometer had not been registered. On the 3rd snowbirds were seen, and marmots had some time before emerged from their winter quarters.

The Esquimaux, with the exception of one or two families, had built their snow huts within a quarter of a mile of our house, where they had been living for more than a week. They had almost all behaved well, and were commended accordingly. They had not yet commenced seal hunting, but were to do so as soon as the seals came up on the ice; in the meantime they were catching deer in snow traps made by digging holes in the snow, and covering them with thin slabs of the same material. Wolves are often taken in a similar manner; but for them the hole requires to be not less than eight or nine feet deep, and after it is covered with a thin plate of hard snow (on the centre of which a bait is laid), a wall is built round it, over which it is necessary for the wolf to leap, before he can reach the bait. He does so, and falls to the bottom of the pit, which is too narrow to give him room to make a spring to the top.

I may now say a few words about our travelling companion Ivitchuk, who had behaved well throughout the journey. We found him always willing and obedient, and generally lively and cheerful except when very tired, which was frequently the case, as he had not been accustomed to travel so many days consecutively. He accommodated himself easily to our manners and customs in every respect, living as we did, though he would swallow a piece of seal's blubber now and then as a delicacy.

What surprised me most was, that he was by no means a very great eater, being often satisfied with as little as any of the party. Tea and chocolate were favorite beverages with him, and he had learned to smoke his pipe as regularly as if he had been accustomed to it all his life. He picked up a few words of English, which he made use of whenever he thought they were applicable, and was very anxious to be taught to read and write. As he, like the rest of the party, was much thinner than when he commenced the journey, he had made up his mind to do nothing during the remainder of the spring but eat, drink, and sleep, a determination to which I believe he most strictly adhered. It was with no small pride that he received a gun and some ammunition, as a reward for his services, and a few presents to his wife, one of the best looking of the fair sex of Repulse Bay, made the pair quite happy, although it was said that the lady had not behaved very well to her liege lord during his absence, having taken unto herself another husband named Ou-plik; but probably the good man knew nothing, or cared little, about it.

Part of the men were now every day occupied in scraping among the snow for moss and heather, of which a sufficient quantity was procured to keep the kettle boiling.

On Sunday the 9th divine service was read, and thanks offered to the Almighty for having guided us in safety through the late journey. Many Esquimaux were present, who conducted themselves with propriety.

CHAPTER SEVEN

Preparations for exploring the coast of Melville Peninsula—Outfit—Leave Fort Hope—Pass over numerous lakes—Guide at fault—Dease Peninsula—Arrive at the sea—Fatigue party sent back to Fort Hope—Barrier of ice—Lefroy Bay—Large island named after the Prince of Wales—Detained by stormy weather—Short allowance—Cape Lady Simpson—Selkirk Bay—Snow knee-deep—Capes Finlayson and Sibbald—Deer shot—A cooking scene—Favourite native relish—Again stopped by stormy weather—Cape McLoughlin—Two men left to hunt and fish—Cape Richardson—Chain of islands—Garry Bay—Prince Albert range of hills—Cape Arrowsmith—Coast much indented—Baker Bay—Provisions fail—Proceed with one man—Cape Crozier—Parry Bay—Cape Ellice, the farthest point seen—Take possession—Commence our return—No provisions procured by the men left behind—Short commons—Flock of cranes—Snow-blindness—Arrive at Repulse Bay.

On the 12th of May preparations were commenced for a journey along the west side of Melville Peninsula. In expectation of falling in with much rough ice, I determined on taking dogs only for the first three days of the journey. The party was to consist of Corrigal (our snow-house builder), Folster, Matheson, and Mineau, with Ouligbuck as deer-hunter and interpreter. A fatigue party of two men, and an Esquimaux with a sledge and good team of dogs, were to accompany us for three days, which I supposed would be the time required to reach the coast.

Our provisions for the journey were two bags of pemmican, each 90 lbs., 70 reindeer tongues weighing nearly 30 lbs., 36 lbs. flour, and a little tea, chocolate, and sugar. We took also a gallon and a half of alcohol and a small quantity of oil.

Leaving George Flett in charge at Fort Hope, we started at 10 P.M. on the 13th of May, and directed our course towards a chain of lakes in nearly a due north direction. Although the snow was soft, and we had some rather steep rising grounds to pass over, we made good progress, and after crossing six small lakes we came to some high table-land, on which the snow was very deep, and in which the sledge sank very much. A walk of four miles brought us to another lake of considerable size. A little after 6 A.M. on the 14th, we found some snow huts that had been inhabited during part of the winter by the Esquimaux Ecouchi, and soon had one of them cleared out for the accommodation of the party.

Although we had not travelled much more than twenty miles, Ouligbuck was so fatigued that I determined to send him back with those who were to return to Repulse Bay. We saw no game and only very few tracks of deer. The weather was so cloudy that no meridian observation of the sun could be obtained. Our latitude was 66° 52' N., and longitude 86° 46' W., both by account.

We resumed our march at 9 P.M. on the 14th, the night being calm, with a little snow falling. A brisk walk of two miles to the N.W. brought us to the end of the lake, when we followed the bed of a small stream to the northward for five miles. Two narrow lakes were next traversed, when our guide, who appeared to know little about the proper route, led us to the N.W.; and after crossing five lakelets, and as many short portages, at half-past 6 A.M. we came to a body of water about the size of that near which we had encamped the day before. Here we stopped for the day. The ice on this lake was six feet thick, and gave the men much trouble to cut through it. There was very little fuel to be found; we were therefore obliged to burn part of the small quantity of oil we had taken with us. By a meridian observation our latitude was 67° 5' 3" N., variation of the compass 53° 30' W., and longitude by account 87° 8' 54" W. The west side of the creek, and also of the lakes which we passed over this day, was steep and rocky, although not high; the east sides were more sloping.

It was near 10 o'clock at night when we commenced our journey. After an hour's walk we came to the north end of the lake, but our young Esquimaux never having been here before (which was rather surprising, as his usual winter home was not more than ten miles distant), was quite at a loss what direction to take. It would have been quite easy for me to have made a straight course by compass, but by doing so we were very likely to get among ground so uneven, as to be impassable to the dogs and sledge. We now turned to the east of north, and after crossing a number of small lakes, arrived at the sea (which here formed a deep inlet) at a few minutes before midnight. Proceeding down the inlet, which for a couple of leagues was not more than half a mile wide, with steep rocky shores (in some places precipitous), we came to rough ice, and found that there were apparently two openings leading to the northward. I chose the one to the left, but we had not gone more than a mile-and-a-half, when we found that we were in an arm of the inlet, and that the land to the north of us, which I had supposed to be an island, was joined to the mainland by an isthmus not more than 50 yards wide. This peninsula I named after P.W. Dease, Esq., the able leader, in conjunction with T. Simpson, of the expeditions which explored so large a portion of the Arctic shores in 1837, 1838, and 1839.

Retracing our steps, we now followed the opening to the right, in which there were great quantities of rough ice, over which we advanced but slowly. The inlet (to which I had given the name of Cameron, after a friend), soon became broader and the ice less rough. At 7 A.M. on the 16th we arrived at the Cape, which last autumn had been named after the late Thomas Simpson, whose agreeable duty it would have been, had he survived, to accomplish the survey which I was now endeavouring to bring to a successful termination. The shores here were very barren, there being little or no vegetation to be seen, except small patches in the crevices of the rocks. In a small lake near our encampment, from which we obtained water, the ice was found to be five feet thick. A sufficient quantity of fuel was gathered to boil our kettle, and two hares were shot by Corrigal. We here made a "cache" of some pemmican, flour, &c. for our return journey. Our snow hut was built on the south side of the cape, under shelter of rocks, near which there were two small islands.

The sledge was to be sent back to Repulse Bay from this place, and with it Ouligbuck, who from his inability to walk would have been an incumbrance to us. The weather was so cloudy that no observation could be obtained. Our latitude by account was 67° 22' (which I afterwards found by observation to be nearly three miles too far north), longitude 87° 3' W. The whole of these three days' journeys had been measured with a well stretched line, but this we could not expect to carry on further, as each person would have enough to do with his load.

Bidding adieu to our companions who were to return to Fort Hope, we commenced our journey at half-past 8 P.M., each of my men being laden with about 70 lbs., whilst I carried my instruments, books, and some other articles, weighing altogether 40 lbs. This was but a light burden for me, but as I had to examine different objects on the route, and also to lead the way, I found it quite enough.

As soon as we had fairly rounded Cape T. Simpson, the coast turned to the eastward, and became indented with narrow but deep inlets, all of which were packed full of rough ice. Walking became most difficult. At one moment we sank nearly waist-deep in snow, at another we were up to our knees in salt water, and then again on a piece of ice so slippery that, with our wet and frozen shoes, it was impossible to keep from falling. Sometimes we had to crawl out of a hole on all fours like some strange-looking quadrupeds; at other times falling backwards we were so hampered by the weight of our loads, that it was impossible to rise without throwing them off, or being assisted by one of our companions. We therefore found it better to follow the shores of the inlets than to cross them, although by doing so we had double the distance to go over. Numerous traces of hares were seen, but we could not afford to lose time in following them.

After passing four inlets having some small islands lying outside of them, we came to a rocky point rather higher than any we had yet met with on this side of the bay. The coast to the eastward of Point Cowie (so named after an old friend) became more level, and instead of granite, was covered with mud, shingle, and fragments of limestone. At half past 3 A.M., all of us being sufficiently tired with our night's work, we built our snow hut and a small kitchen for cooking. This was our usual practice when we had found, or were likely to find, fuel. In the present instance, we had the good fortune to collect enough to boil a kettle of chocolate, and we consequently enjoyed an excellent supper, if I may so term a meal taken about six in the morning.

The weather had been fine until midnight, when it began to snow and drift, with a strong breeze from the north. Thermometer +13°. At noon the sky was too much overcast to obtain an observation. Our latitude was 67° 24' 20" N., longitude 86° 37' W. both by account.

When we resumed our journey, at 7 o'clock in the evening of the 17th, there was still a strong breeze from N.N.W. with snow drift, the temperature being +18°. Our snow hut of the previous day we now found to be on the shore of a large bay, the most distant point of which bore nearly due north. To follow the coast would have cost us a great deal of additional walking; I therefore determined to attempt the traverse of the bay towards the point above referred to. All along the coast there was a belt of rough ice about two miles broad, over which we were forced to pass before reaching some that appeared smoother outside. To cross this barrier occupied us more than two hours, and gave us more violent exercise than all the remainder of the day's journey. It was half-past 3 A.M. when we arrived at the north point of the bay, which was low and level, with some hills a few hundred feet high, three or four miles inland. We had passed two small rocky islands to seaward in the first part of the night, and there was another close to a bluff point on the south side of the bay. To this cape I gave the name of Watt. The bay was called after Lieut. (now Captain) Lefroy of the Royal Artillery, whose name is well known to the scientific world, and of whose kindness in aiding me in my astronomical studies I retain a most grateful remembrance.

We crossed over to Cape W. Mactavish (so named after William Mactavish, Esquire, chief trader, an intimate friend, to whom I am much indebted for assisting me in fitting out the expedition,) and stopped about three miles beyond it. Here we built our snow hut, which was found by meridian observation to be in latitude 67° 42' 22" N.;

the variation of the compass 80° 35' W., and the longitude by account 86° 30' W. Directly opposite our encampment, and extending for about seventeen miles to the northward of it, there was a large island of table land, with not a single rock *in situ* to be seen on it. Its southern extremity bore nearly west (true) from us, and the strait which separated it from the mainland was not more than a mile and a half wide. This island was honoured with the name of His Royal Highness the Prince of Wales, and a smaller one to the south of it was named after Colonel Sabine.

Not a single living animal had been seen all day, but some traces of deer proceeding northward were noticed. We were again fortunate enough to find a little fuel.

Our route on the following night was nearly straight in a N.N.E. direction. The snow was very soft and deep in many places. A few hundred yards from the beach there were steep banks covered with shingle and small boulders of granite, where we usually found the snow less deep, and walking consequently better. After travelling nine miles we came to a considerable creek, about twenty yards wide, in which a deep channel had been worn among the mud and shingle. Near it there were numerous Esquimaux marks set up, and circular tent sites, but all of old date. We continued our march twelve miles further, and at 8 A.M. arrived at another creek somewhat larger than the last, and with higher banks. Here there were also many Esquimaux marks, and I afterwards learned that some parties had resorted hither from Repulse Bay, for the purpose of catching salmon, trout, &c. About an hour before reaching this place we crossed a long and curiously shaped point, which I named Point Hamilton after a near relative. The bay formed by it was called Erlandson.

One of the men, although an able active fellow, not being used to this sort of exercise, was much fatigued; and as the weather looked threatening, I ordered our snow-house to be built—the more readily as there was fuel to be found. In little more than a hour and a half we were comfortably housed, and not long afterwards we had taken our usual morning meal of pemmican seasoned with a handful of flour, those forming, when boiled together, a very nourishing and not unpalatable dish. The temperature all night had been 22° above zero, being too warm for walking pleasantly; and the men, having had to exert themselves much, were glad to get to rest as soon as possible, whilst I remained up to obtain a meridian observation of the sun. This gave latitude 67° 58' 49" N. Our longitude by account was 85° 59' 36" W. The sun was too much obscured by clouds to obtain the variation. We here deposited some pemmican and a little flour for our return journey.

When we started at 8h. 30m. P.M. on the 19th it blew a gale of wind from S.S.E. with much drift and snow, the temperature being only 4° below the freezing point. Fortunately the wind was on our backs; but the drift was so thick that we were obliged to follow every turn of the coast, and we could not see more than twenty yards before us. When we had travelled six miles we came to a bay a mile and a half wide, on the north shore of which there were two strangely shaped rocks of granite, having the appearance of an old ruin or portion of a fortress. They were of a square form, each about twenty-five feet high and nearly as much in extent.

Our course now lay due north; but we had not gone more than twelve miles altogether, when the weather became so unpleasant that we were glad to get under shelter, and before we did so, every part of our clothes was penetrated with snow drift. We could obtain no fuel here.

The weather continued so stormy that we were unable to leave our snow-hut until a quarter past 8 P.M. on the 21st. During our detention, finding that our provisions would run short if the walking continued as difficult as it had been, we took only one not overabundant meal during the twenty-four hours. There was still some snow falling, so that I could not take the proper bearings of the land along which we passed. The land, after we had proceeded N.E. for a few miles, turned to the southward of east, forming a bay eight miles wide, which, as it was full of rough ice, we were under the necessity of coasting. This bay was called after the Rt Hon. the Earl of Selkirk, and the cape forming its western boundary was named after the amiable lady of our much respected governor, Sir George Simpson.

The snow was in many places so soft and deep that we sank above the knee at every step, which made our night's march fatiguing in the extreme. On the N.E. side of Selkirk Bay, which is steep and rocky, there was a deep indentation or inlet, into which two small creeks emptied themselves. The land for five miles had a N.W. trending, and again turned up to the eastward of N., forming a high rugged headland, which was named Cape Finlayson, after Duncan Finlayson, Esq., Chief Factor. At three miles from Cape Finlayson we passed Point Barnston, and about four miles beyond this we came to another rocky point, which received the name of Cape Sibbald. The night had now become very disagreeable, with a heavy fall of snow; we persevered notwithstanding, partly crossing and partly coasting a bay heaped with rough ice, and encamped on what I supposed was its northern extremity, but which afterwards turned out to be an island, and to which I gave the name of Glen. The bay we had just passed was called after William G. Smith, Esq., Assistant Secretary to the Hudson's Bay Company.

The snow not being in a good state for building, we were rather longer than usual in getting housed. There was no fuel to be found, so we followed our old plan, and took a kettle or two of snow to bed with us. The temperature was very high for the season, being only 5° below the freezing point.

When we started at a quarter-past 11 on the 22d, the night was beautifully clear and calm, with the thermometer at 13° below zero. After a three hours' walk we arrived at the north point of a bay, three and a half miles wide, across which we had come. To the bay I gave the name of Fraser, and to the point that of Corcoran, after two intimate friends, chief traders of the Company.

We had not advanced many miles farther, when some deer were noticed at no great distance, feeding on the banks of a stream. Being desirous of procuring some venison if possible, I sent Corrigal (who, with other good qualities, was a very fair shot) after them, and he was fortunate enough to shoot a fine buck. But the buck, though wounded, could still run too fast to be overtaken, and the sportsman was just about to give up the chase when I joined him, and we continued the pursuit together. The deer, having got a considerable way in advance, had lain down, but rose up before we could get within good shooting distance, and was trotting off at a great pace, when, by way of giving him a parting salute, I fired, and very luckily sent a ball through his head, which dropped him. His horns were already about a foot long, and the venison was in fine order for the season of the year.

I immediately returned to the men, who had been busily employed collecting fuel, of which great quantities grew along the borders of the creek, and sent two of them to assist in skinning and cutting up the deer, whilst I and the other men continued to gather heather, as we now anticipated great doings in the kitchen. We placed the greater part of our venison "en cache," but kept the head, blood, leg bones, &c., for present use; and being determined to lose nothing, the stomach was partially cleaned by rubbing it with snow, and then cut up and boiled, which thus made a very pleasant soup, there being enough of the vegetable contents of the paunch to give it a fine green colour, although I must confess that, to my taste, this did not add to the flavour. Having discussed this mess, a second kettle full was prepared, composed of the blood, brains, and some scraps of the meat, which completed our supper.

It is well known that both Esquimaux and Indians are very fond of the contents of the paunch of the rein-deer, particularly in the spring, when the vegetable substances on which the animal feeds are said to be sweeter tasted. I have often seen our hunter, Nibitabo, when he had shot a deer, cut open the stomach, and sup the contents with as much relish as a London alderman would a plate of turtle soup.

The position of our snow-house was in latitude 68° 33' 26" N., longitude 85° 20' 30" W., both by account.

The weather was so stormy during the 23rd that we could not continue our journey. The thermometer rose as high as +39° in the shade, and the melting of the snow having wet the heather, we were obliged to have recourse to alcohol. Three or four snow buntings and traces of partridges (*tetrao rupestris*) were seen.

On the 24th it still blew a gale of wind from the east, but there being a partial thaw by the high temperature, there was no drift, and much of the ground was entirely cleared of snow.

In the evening the weather became more moderate, and the thermometer fell to 5° below the freezing point. We started at a few minutes after 10 o'clock, our course being slightly to the east of north. The travelling was still very fatiguing, as we were frequently forced to pass over the rocks, or to walk along the steep drift banks, in order to avoid the rough ice which had been heaped up against the shore. We passed a number of small bays and points, and when we had advanced fifteen miles, came to a high cape, which forms the N.W. promontory of a bay five miles in extent. To the cape I gave the name of McLoughlin, after the gentleman who has been for many years in charge of the Columbia department, and the bay was called after my much valued friend Nicol Finlayson, Esq., Chief Factor. After passing Cape McLoughlin we turned to the eastward, toward the head of the bay, and stopped at 7 A.M. near the mouth of a creek, where we took up our quarters for the day.

There was not so much fuel to be found as at our last encampment, but we gathered enough to boil our kettle. Some bands of deer and a few partridges were observed, but we did not waste time in endeavouring to get a shot at them. Since leaving Fort Hope not a day had passed without more or less snow falling, which made the travelling much more difficult than I expected, and our progress consequently so much slower, that, notwithstanding the addition I had made to our stock of provisions, there was some danger of our still running short. I therefore decided on leaving two of the men here to fish and shoot, whilst I went forward with the others.

There was a little snow falling when, along with Corrigal and Matheson, I set out at 10 P.M. on the 25th. The night was mild (6° below freezing) with a light wind from the east. A walk of two miles brought us to a head land, which formed the north side of Finlayson Bay, and which extended seven miles in a W.N.W. direction. To this cape the name of Richardson was given, after the distinguished naturalist, who, having already exposed himself to many dangers and privations in the cause of science, is now about to incur similar hardships in the cause of humanity and friendship, by searching for Sir John Franklin and his gallant party, whose situation, it is too much to be feared, is a critical one.

At the place where we crossed Cape Richardson it was not more than a mile wide, and we found ourselves in a large bay, thickly studded with high and rugged islands. A chain of these islands, which lay outside of us, and to which I gave the name of Pomona, (after the largest island of the Orcadian group,) had effectually served as a barrier to the ice from seaward, and had thus made the walking much smoother than we had hoped to find it. As we advanced there were many tracks of polar bears, and also those of a wolverine, that appeared to follow them very closely, expecting no doubt to appropriate some portion of whatever prey they might catch. A flock of long-tailed ducks passed us, flying to the westward, towards some open water, the vapour exhaled from which appeared in that direction.

As we approached the north side of the bay, which was named after Nicholas Garry, Esq., of the Hudson's Bay Company, there were so many islands that I was much at a loss what direction to take. Under these circumstances we encamped at 6 A.M. on a high island, about two miles in diameter, from which a good view could be obtained. Garry Bay is the most strangely shaped, and the most irregular in its outline, of any we had yet seen. It presented three long, narrow, and high points of land, and had four inlets. The largest and most southerly of these points was called after Lieut. Halkett, R.N., and the most northerly of the inlets received the name of Black Inlet. As no fuel could be obtained here, we were reduced to the necessity of using some more of our alcohol, of which but a small quantity now remained. The men were soon asleep under our single blanket, (for this was all the covering we had for the party,) whilst I remained awake for the purpose of obtaining an observation of the sun at noon. This gave latitude 68° 59' 15" N., variation of the compass 88° 26' W., our longitude by account being 84° 48' W.

All the way between Lefroy and Garry Bays there is a range of hills, from 500 to 800 feet high, about five miles from the coast, which was distinguished by the name of His Royal Highness Prince Albert, consort of our beloved Sovereign.

The weather was beautiful all day, and was equally fine when we commenced our march at half-past nine at night. Our route lay somewhat to the west of north, between two lofty islands, the smaller of which received the name of Gladman, and the larger and most northerly I designated Honeyman, after a brother. Seven miles from our encampment we passed a bluff and precipitous point, the northern extremity of Garry Bay, to which the name of Cape Arrowsmith was given, in honour of John Arrowsmith, Esq., the talented hydrographer to Her Majesty.

The land was now completely serrated with narrow points and inlets, along which we were able to make nearly a straight course, as the force of the ice from the westward had been much broken by ridges of rocks that lay outside of us. To four of these inlets I gave the names of McKenzie, Whiffen, Bunn, and Hopkins, after much esteemed friends.

Towards the end of our night's journey the coast turned nearly due north, and when we had advanced seven leagues we encamped on Cape Miles,—so named after Robert Miles, Esq., Chief Factor,—at 7 A.M. on the 27th. As the morning was exceedingly fine, we thought there was no necessity for building a snow-house, an omission which we regretted in the afternoon, when a heavy fall of snow took place.

By a good meridian observation of the sun, the latitude 69° 19' 39" N., and the variation of the compass 92° 20' west, were obtained, the longitude by account being 85° 4' W. The latter is evidently erroneous, as I had neither chronometer nor watch that I could place dependence upon, and the compasses were much affected by local attraction.

Our provisions being now nearly all used, I could advance only half a night's journey further to the northward, and return the following morning to our present quarters. Leaving one of the men, I set out with the other at half-past 9 P.M., the snow falling fast; and although we had little or nothing to carry, the travelling was very fatiguing as we crossed Baker Bay—so named in memory of a much valued friend—at the north side of which we arrived after a walk of four miles. It now snowed so thick that we could not see farther than fifty yards round us, and we were consequently obliged to follow the windings of the shore, which, when we had traced it six miles beyond Baker Bay, turned sharp to the eastward; but the weather continuing thick, I could not see how far it preserved this trending. After waiting here nearly an hour, the sky cleared up for a few minutes at 4 A.M., which enabled me to discover that we were on the south shore of a considerable bay, and I could also obtain a distinct view of the coast line for nearly twelve miles beyond it.

To the most distant visible point (latitude 69° 42' N., longitude 85° 8' W.,) I gave the name of Cape Ellice, after Edward Ellice, Esq. M.P., one of the Directors of the Company; the bay to the northward, and the headland on which we stood, were respectively named after the distinguished navigators Sir Edward Parry and Captain Crozier.

Finding it hopeless to attempt reaching the strait of the Fury and Hecla, from which Cape Ellice could not be more than ten miles distant, we took possession of our discoveries with the usual formalities, and retraced our steps, arriving at our encampment of the previous day at half-past 8 A.M. Here we found that Matheson, the man left behind, had built a snow-house after a fashion of his own, the walls being like those of a stone building, and the roof covered in the same way with slabs of snow placed on the opposite walls in a slanting position, so as to rest on one another in the centre. Seven hours had been spent in building this edifice, which was not a very handsome one; but being sufficiently wide, and, when our legs were doubled up a little, long enough for us all when lying down, we found it pretty comfortable.

During the remaining four hours of our absence, he had been engaged in an attempt to coax a little wet moss into a sufficient blaze to boil some chocolate; but, notwithstanding his most persevering exertions, by the time his fuel was expended, the chocolate was little more than lukewarm, although our cook *pro tempore*, who was of a sanguine temperament, firmly believed that it was just about to reach the boiling point. We finished the process with a little of our remaining stock of alcohol, and enjoyed an excellent though rather scanty supper.

Matheson was one of the best men I ever had under my command.

Always ready, willing, and obedient, he did his duty in every respect, and whilst he possessed spirit enough for anything, he had a stock of good humour which never failed him in any situation, however difficult and trying. Were the walking difficult or easy, the loads heavy or light, provisions abundant or reduced to less than half allowance, it was all one to Peter Matheson; he had a joke ready for every occasion.

A few minutes after 10 P.M. on the 28th, we were on the march homeward. The night was very disagreeable, there being a strong breeze of head wind with heavy snow, and a temperature much too mild (only 8° below the freezing point) for walking comfortably. The snow also was very soft, so that, had it not been for the bad state of our victualling department, we would have remained snug in our quarters. But needs must when hunger drives, so we trudged on stoutly, crossing over the land for the purpose of shortening our distance. After a tough walk, during which we met with some tracks of bears that had passed only about an hour before, we encamped on a small island close to Cape Arrowsmith, and nearly three miles to the northward of our snow hut of the 26th. The weather during the day became fine, so fine indeed that our house, not being built of good material, tumbled down about our ears just before we were leaving it.

29th.—When we started at half-past 9 P.M., the night was fine, but in half an hour it began to snow so thick that we could not keep our course in crossing Garry Bay, where the walking was much worse than when we formerly passed. In three hours the weather again cleared up, and I found that we had not deviated much from the right road.

At 7 A.M. we joined Folster and Mineau, whom we found quite well, but like ourselves very thin. The only animals they had killed were two marmots, and no fish had been caught. If we had been twelve hours longer absent, they intended to have boiled a piece of parchment skin for supper, and to have kept the small remaining piece of pemmican for travelling provisions.

I have had considerable practice in walking, and have often accomplished between forty and fifty, and, on one occasion, sixty-five miles in a day on snow shoes, with a day's provisions, blanket, axe, &c. on my back; but our journey hitherto had been the most fatiguing I had ever experienced. The severe exercise, with a limited allowance of food, had much reduced the whole party, yet we were all in excellent health; and although we lost flesh, we kept up our spirits, and marched merrily on, tightening our belts—mine came in six inches—and feasting our imaginations on full allowance when we arrived at Fort Hope.

On the 30th we continued our course homewards, crossing over the several points that we had formerly coasted. It snowed heavily all night, and the temperature was only two degrees below the freezing point. Eight cranes "winged their circling flight" northward, and half a dozen sandpipers were seen. It was near 4 A.M. on the 31st when we arrived at our snow house of the 23rd, which we found quite as good as when we left it and our cache of venison all safe. Three partridges were shot, which somewhat aided our short commons.

On the following night, after an ineffectual attempt to get to seaward of the rough ice, in which we lost a considerable portion of the skin off our shins, we travelled on the land, making short cuts whenever practicable.

On arriving opposite to Glen Island, we found that it was divided from the shore by a channel not much more than a quarter of a mile wide. There was an inlet a few miles in length to the eastward of it, which was named after the Rev. Mr. Mackar of Kingston, Canada West. This night was the finest we had experienced throughout the journey.

A specimen of trap rock was obtained from some rising grounds a mile and a half distant from the north shore of Smith's Bay, near the head of which we now for the first time observed a lake of a couple miles in extent. When half a league from Cape Sibbald, we encamped under shelter of some precipitous trap cliffs nearly a hundred feet high. Some more cranes were seen, and numerous traces of deer and partridges. We here procured some fuel, there being patches of ground bare of snow. Our latitude by observation was 68° 19' 50" N. Variation of the compass 80° 55' W. Two of the men were affected with snow blindness—one of them severely.

1st June.—It blew a gale of wind from S.E., with thick snowdrift at 8h. 30m. P.M. when we resumed our journey. At half-past 10 we crossed the largest stream that we had yet met with on Melville Peninsula. It was already partially open, owing to numerous springs, which had formed many small mounds of ice from ten to twelve feet high. After taking a copious draught from the limpid stream, we continued our journey across Point Barnston and Cape Finlayson, until we arrived at Selkirk Bay, when, the weather having become much worse, we stopped at 1h. 30m. A.M. to build our snow hut at a place where there was such an abundant supply of heather, that we had enough to cover our snow-bed with. Two deer were seen, and Corrigal made an ineffectual attempt to get a shot at them. I shot five ptarmigan, and four sandpipers were observed.

During the next night's journey the weather was very snowy, but the wind being more moderate we got on faster. After coasting Selkirk Bay, we cut across Cape Lady Simpson, and at half-past 6 A.M. on the 3rd of June, we reached our encampment of the 19th ultimo in Erlandson Bay, where we found our small "cache" of provisions quite safe. Five more partridges were shot, and some deer seen. The snow being very soft, we remained here all day, and at noon obtained the latitude 67° 59' N., and variation 75° 9' W. The thermometer in the shade rose as high as +54°, and our old snow-house tumbled down about our ears in the evening, just as we were going to take our supper,—perhaps breakfast would be the more appropriate term, as we had turned day into night.

We started at 8h. 30m. P.M., and notwithstanding the great power of the sun, so much snow had fallen lately that it lay far deeper on the ground than when we had previously passed this way. The walking also was so much more fatiguing, that we were not able to reach our snow-house of the 18th of May, and were in consequence under the necessity of building new lodgings. The night was mild and nearly calm. Two phalaropes (*P. fulicarius*) were seen, and a couple of ptarmigan shot. There was no fuel to be found here, but having picked up a little as we came along, we did not feel the want of it much.

The 4th was a fine night with the thermometer at +23°, when, at 7h. 40m., we resumed our march. Whilst rounding Cape Mactavish we fell in with nine partridges, seven of which were shot, and I endeavoured to get within range of a couple of swans—the first we had seen—but they were too shy. We now crossed Lefroy Bay, the snow on which was very soft, and built our snow-house on the ice at 7h. A.M. about four miles from its south shore. The work during this journey had been so much more severe than was expected, and the men had in consequence used so much more tobacco than they had anticipated, that their stock was now quite exhausted, and they appeared to feel the want as much as if they had been deprived of half their allowance of food,—perhaps more. It was really amusing to see how very particular they were in dividing the small remaining bits which they rummaged from the dust and rubbish in their pockets, and which at any other time they would have thrown away. I happened to have a little snuff with me, a pinch of which, in their necessity, they relished much.

We were on foot again at 20 minutes after 8 on the 5th. The weather had been stormy all day, but became fine an hour after we started. We kept well out from land, expecting to find the ice smoother; and this was the case as far as Point Cowie; but beyond that the rough ice extended quite across the bay; we therefore struck in for the shore, which after two hours' scrambling we reached, and directed our course over the rocks,—from which the snow had now, in many places, entirely disappeared,—towards Cape T. Simpson, where we arrived at 5h. A.M. on the 6th, and found our "cache" of provisions, &c., as we had left it. No time was lost in getting the stones cleared away from it, not so much for the purpose of having something to eat, as to find some tobacco that had been left here among other things. A fine hare had been shot, and as soon as three of the party, who had stopped behind to gather fuel, came up, we had a much more abundant and palatable meal than we had enjoyed for many days before. To the large bay, the survey of which we had now completed, the name of Committee Bay was given, in honor of the Committee of the Hudson's Bay Company. This was the finest day we had experienced during this journey, the power of the sun being so great as to raise the thermometer to +82°.

By an excellent meridian observation in quicksilver, our latitude was 67° 19' 14" N., variation of compass, 64° 27' W. Wishing to take a straighter, and consequently shorter, route to Repulse Bay than that by which we had gone, we started at 9 P.M. on the 6th, and after a walk of three hours came to the head of a narrow inlet, with high rocky shores, and about seven miles long, to which I gave the name of Munro. Our course overland was nearly due south, and we passed over a number of small lakes, from which the snow had been partially removed by the joint action of the sun's rays and the wind.

On the following night our course continued the same with a slight inclination to the westward. We had a strong gale of fair wind, which helped us along amazingly; but as we could easily reach Fort Hope in another night, and as we had abundance of food, we encamped at 3h. 30m. A.M. on the 8th, during the whole of which day, until late in the evening, it blew hard with drifting snow, so that no observations could be made.

Being anxious to arrive at winter quarters early on the following day, we were again on the march at half-past 7 P.M., and the evening having now become fine, we kept up a smart pace for a few hours until we arrived at Christie Lake, where, finding some very fine heather quite dry and free from snow, it was impossible to resist the temptation of having something to eat and drink. Having taken up our quarters in an old snow-hut, the chocolate and pemmican kettles were soon on the fire, and we heartily enjoyed our rather unusual meal. Following the lake and North Pole River, we came to Fort Hope at 8h. 20m. A.M. on the 9th, all in good health and spirits, but very much reduced in flesh, although not quite so black as when we returned from the previous journey.

CHAPTER EIGHT

Occurrences at Fort Hope during the absence of the exploring party—Remove from winter quarters to tents—Sun seen at midnight—Build an oven and bake bread—Esquimaux method of catching seals—A concert—Lateness of the summer—A native salmon-wear—Salmon spear—Boulders on the surface of the ice—Visited by a native from the Ooglit Islands—His report of occurrences at Igloolik—Indolence of the natives—Ice breaking up—Halkett's air-boat—A storm—The ice dispersed—Prepare for sea.

During my absence from Fort Hope little beyond the usual occurrences of the winter had taken place. The latter part of May was remarkable for the great quantity of snow that fell, with gales of wind and drift, which kept the men almost continually clearing away snow from the roofs of our houses. They were obliged even to go to work during the night, and notwithstanding all the care that was taken, two of the boats' yards were broken, and the masts very nearly shared a like fate, as the post placed under them gave way. For so great a quantity of snow lodging on our roof, the man left in charge was to blame, as shortly after my departure he had the snow thrown up in heaps, which, when the stormy weather and snowdrift came on, caused drift-banks to be raised to an equal height (about 4½ feet) on the tops of our dwellings.

During all this time the thermometer never fell lower than +9°, which was on the 16th of May, and rose as high as +45°, at mid-day on the 29th. The last day of May was very stormy; but on the 1st of June the weather changed for the better, although the thermometer was as low as +12°. On this day the first geese (laughing geese) and some sandpipers were seen, and one of each was shot. As the partridges were migrating northward about thirty had been killed, and there was a good stock of venison in store, the hunters having shot twenty deer. The does were now very large with young, and had become very poor; the bucks were, however, improving in condition.

The Esquimaux had brought in little for trade, a few pairs of boots, which were soon bought up by the men, and a little oil from Akkeeoulik being the principal articles. Some of them were getting short of provisions, not having been able to find a "cache" which they went for. They had all behaved well, not having committed any thefts that could be discovered. We had, however, one most incorrigible thief among our party, Ouligbuck's son, who, during the few days of his father's absence, was twice caught with the old man's bale open, eating sugar; some tobacco was also taken, and the trousers of most of the men were completely cleared of buttons by the same hands. On my return only one family of Esquimaux (Shimakuk's) remained near us. Shimakuk had been waiting for his dogs, which were with the party who had gone in search of meat.

On the 13th divine service was read, and thanks returned to the Almighty for His protection throughout the winter and during the late journey.

There was a strong breeze of N. wind, with frequent showers of snow. House very damp; the clay falling from the inside of the walls.

14th.—The weather was fine and permitted us to get our flour, pemmican, &c., removed from the meat store (which was now dropping much from the roof) to the rocks, where it was well covered up with oilcloths.

The 20th was a most stormy day with occasional showers—wind N.W. There was a considerable stream of water running on the ice of North Pole River, forming large pools on the sea-ice, through which it did not yet find a free exit.

21st.—There was a change in the weather for the better, although it still blew a gale; however, as the day advanced the wind became more moderate, and about noon shifted round to the south.

The water was rising fast in all the creeks, showing that the process of destruction was fast going on among the snow and ice. The latter was still nearly four feet thick on the lakes, but very porous.

The great rise of water in the creeks and small streams rendered it very unpleasant and even dangerous to cross them. In attempting to get near some geese this day I sunk to the waist amidst snow and water, and not being able to get any firm footing, I found much difficulty in scrambling out without wetting my gun.

23rd.—This being a fine day, all the men were employed dismantling the house and carrying down the provisions, clothes, &c. to the summer tents, which had been pitched about 300 yards nearer the shore. Two leather tents were put up for cooking in. We saw the sun at midnight, his lower limb touching the high grounds to the northward.

We made some bread in an oven which we had built of stones cemented with clay of an excellent quality. The upper part of our first batch was well baked, but the floor of the oven was not sufficiently warm to bake the lower part. It however rose well, and we afterwards succeeded in making excellent bread, though the oven was heated with heather.[4]

15th July.—Weather still stormy and cold to the feelings, the thermometer being +35°. The water of North Pole Lake had broken through its barrier of snow and ice, and was rushing down the river with great force, carrying with it large masses of ice.

All the men except Flett, who remained at the tents, and Germain, who had charge of the nets, went to North Pole Lake on the 19th to bring down the boat. The river being one continued rapid throughout its whole length, with not an eddy to stop in, they came down at rather a quick rate, but were compelled to stop within a few hundred yards of the salt water, on account of the shallowness and the number of stones. Twenty-two salmon were caught, some in good condition, others very soft and thin. The former contained roe about the eighth of an inch in diameter.

A number of Esquimaux arrived for the purpose of catching salmon, having finished their seal hunting, which had been successful, although the number killed could not be ascertained. Our old friends were accompanied by three strangers, viz., an old man and two young ones, with their wives and families. Our travelling companion Ivitchuk had shot some deer with his gun, but having spent nearly all his ammunition, he requested and obtained a small additional stock.

Another Esquimaux, a jolly old fellow, with two wives, joined the party here; he had come from the direction of Wager River this spring on the ice. He and one or two more old men were nearly starved to death last winter, being so much reduced that they could not walk. Twenty-three salmon were got from the nets; some of these were in very poor condition, being evidently out of season; others were in fine order and full of roe.

22d.—One of the old Esquimaux at the fishery speared a seal on the ice near the edge of the open water, but it got away in consequence of the line breaking. Their mode of approaching the seal requires much patience and is very fatiguing, as the hunter must lie flat on his face or on his side, and advance towards the seal by a series of motions resembling those of the animal itself. He has frequently to proceed in this way some hundred yards but so well does he act his part that he can get within a few feet of his object, and a looker-on would find much difficulty in telling which was the man and which the seal.

The seal actually comes to meet the hunter, who, as soon as it has got some distance from its hole, springs up and intercepts its return. The women are very expert at this mode of hunting, and frequently having no spear, use a small club of wood with which they strike the seal on the nose.

The greater part of the Esquimaux were encamped about a quarter of a mile from us, and had a *concert* every night,—a union of the vocal and the instrumental. Their only musical instrument is a sort of drum or tambourine, consisting of a stout wooden hoop, about 30 inches in diameter, round which, when it is to be used, a wet parchment deer skin is stretched. In beating this rough instrument, the hoop, not the skin, is struck. The performer being in the centre of the tent, keeps turning slowly round, whilst four or five women add their voices to the execrable sound, producing among them most horrible discord. Each of the men in his turn takes up the drum and thumps away till he is tired, when he lays it down and another takes his place, and so on it goes until it has passed through the hands of all the males of the party, including the boys.

The whole of the natives, with the exception of a few old people who were remaining at the fishing station, and three young men and their wives, went the following day to an island four miles off for the purpose of killing more seals, and also to put new covers on their canoe frames.

25th.—This was the anniversary of our arrival here last year; and certainly everything wore a very different aspect from what it then did. Last summer at this date there was no ice to be seen in Repulse Bay; the snow had nearly all disappeared, and the various streams had shrunk to their lowest level. *Now* there was not a pool of water in the bay, except where the entrance of a river or creek had worn away or broken up the ice for a short distance. There was much snow on the ground in many places, and most of the streams were still deep and rapid.

The musquitoes were rather troublesome, but this I was not sorry for, as the Esquimaux said that the ice in the bay would soon break up after these tormentors made their appearance.

As our native friends were now getting sufficient fish to maintain them, they required no further assistance from us at present. Their mode of catching salmon is a very simple one. They build a barrier of stones about 1½ or 2 feet high across a creek, some distance below high-water mark. The salmon, which keep close to the shore at this season, are by this means, during the ebb of the tide, cut off from the sea, and are easily speared. About sixty were thus killed this day. The spear used is usually made of two diverging pieces of musk-ox horn, from 4 to 5 inches apart at the extremities; between these there is a prong of bone about 3 or 4 inches shorter than the outer ones. Each of the longer prongs is furnished with a barb on its inner side, made of a bent nail or piece of bone, which prevents the fish from escaping. The handle is 6 or 8 feet long. The head of the instrument much resembles a three-pronged fork, with the middle prong a little shorter than the others.

The moon was full this day. High water at 45 minutes past noon. Arkshuk, Shimakuk, and Kei-ik-too-oo visited us on the 28th, bringing a few pairs of boots for sale. The tins which contained preserved meat, and table knives and forks, were in great demand among these good folks. One of the ladies to whom I gave a fork, used it as neatly in eating fish as if she had been accustomed to it from childhood. Thermometer as high as +60° in the shade.

The ice in the bay had broken up for more than a mile from the shore opposite the mouth of the river, but some distance out it looked as white and firm as ever.

I had for some time observed that large stones, some of them of one or two tons weight, were making their appearance on the ice, and I was much puzzled to make out how they came there. They could not have fallen from the shore, as the beach was sloping at the place, nor had they been carried in by drift ice of the previous season. The only way that I could account for it was this. At the commencement of winter the ice, after acquiring considerable thickness, had become frozen to the stones lying on the bottom, and raised them up when the tide came in. The stones would get gradually enclosed in the ice as it grew thicker by repeated freezings, whilst by the process of evaporation, which goes on very rapidly in the spring, the upper surface was continually wasting away, so that in June and July there was little of the first formed ice remaining, and thus the stones which at first were on the under surface of the ice appeared on the top. This may perhaps in some measure account for boulders, sand, shells, &c. being sometimes found where geologists fancy they ought not to be. Ice has been time out of mind the great "conveyancer."

August 1st.—We were visited this day by an Esquimaux named I-ik-tu-ang, whom I had not before seen. He had passed the winter near the

Ooglit Islands, a few days' journey from Igloolik. He said that, when a boy, he was frequently on board the Fury and Hecla in 1822–23, and that the "Kabloonans" killed a number of walruses, and some black whales, with two small boats; that the walruses were put in "cache" for them (the Esquimaux), who were rather short of provisions at the time, and that they received the *skins* of the whales. They had abundance of provisions last winter, but were visited by a very fatal disease—from what I learnt of the symptoms, resembling influenza—which carried off twenty-one grown-up persons. The children were not attacked with this complaint. Two of the party at Igloolik had been reduced to the necessity of putting to death and eating two children, to save themselves from starvation.

Four men, whilst hunting the sea-horse with their canoes lashed together, were assaulted by this fierce animal, struck down with his formidable tusks, their canoes capsized and broken, and the whole party drowned. Another poor fellow having early in the winter harpooned a walrus through a hole in the ice, was dragged into the water before he could disengage himself from the line. The ice being still thin and transparent, the body was found a few days after.

I-ik-tu-ang also informed me—as I had already supposed from various appearances—that there is open water throughout the winter between this and the Frozen Strait, through which a strong current runs with the flow and ebb of the tide,—so strong is it that when bears are pursued and take the water, they are often swept under the ice and drowned.

In the afternoon two more Esquimaux with their wives from the same quarter, accompanied by Akkee-ou-lik and his family, made their appearance.

Some of the natives who had taken up their quarters near us were supplied daily with fish. They appeared quite as indolent as most of the other savage tribes of America, and never thought of looking out for food, so long as they could get enough to support life from us. Although they had a wear made for confining the salmon, they would not take the trouble to spear them when in it.

We endeavoured to get some young marmots, but without success.

I find that these curious little animals leave their winter habitations, which are usually formed in dry sandy banks, as soon as the snow has in a great measure disappeared, and take up their summer residence among the rocks, where, I have no doubt, they are much safer from their numerous enemies.

The weather was still fine on the 6th, but it appeared to have little effect on the ice in the bay, which still remained hard and fast. All the largest and deepest lakes were covered with strong ice.

9th.—On looking out this morning I was happy to see a lane of open water stretching completely across the bay, but there was still a strong barrier between us and the south point, although a passage to the northward might easily have been made. The nets produced eighty salmon, the greater part of which were given to the Esquimaux. The fishery was now abandoned, as we could procure close at hand as many salmon as we required.

During the whole of our spring fishing Halkett's air-boat was used for setting and examining the nets, and was preferred by the fishermen to the large canvas canoe, as it was much lighter, and passed over and round the nets with more facility. Notwithstanding its continued use on a rocky shore, it never required the slightest repair. It is altogether a most useful little vessel, and, as I have said before, ought to form part of the equipment of all surveying parties, whether by land or sea.

The men from the fishery were followed soon after by the Esquimaux with their baggage, which it took more than a dozen trips of our canoe to ferry over.

The large lakes were still covered with a thick coat of ice. There were a great many seals in the open water, and some of the fish in the nets had been eaten by them.

10th.—A storm from the north with rain and snow until noon, when the wind somewhat abated, and the weather cleared up. Great havoc was made among the ice, and in the evening there was a clear sea as far as the point of the bay.

11th.—There was a gale of wind all day with rain occasionally—the weather cold and unpleasant. We were all busily employed in preparing for sea. All the snow-banks for six or eight feet from the ground having been converted into solid ice soon after the spring thaw commenced, we had to dig out the chain and anchor of one of the boats, which were buried under ice of that thickness; yet on the very spot where this chain and anchor lay, there was not a particle of either ice or snow on the 25th July last year; such is the variable nature of this northern climate.

In the afternoon Nibitabo was sent to endeavour to get some fresh venison for our voyage, and shot two young deer; St. Germain and Mineau set the nets for a supply of salmon, and I was busy distributing among the Esquimaux axes, files, knives, scissors, &c. &c. &c.

The large lakes were still covered with ice, but in the bay there was little or none to be seen.

CHAPTER NINE

Voyage from Repulse Bay to York Factory.

Having got every thing ready, the boat launched and loaded about 2 o'clock P.M. on the 12th of August, I was about to distribute our spare kettles, some hoop iron, &c. among the Esquimaux, when the compass of one of the boats was missing. Search was made, but no compass was to be found. At last I thought of turning over some heather that lay close to where my tent had been, and there discovered it. It had been concealed by one of the Esquimaux women—a widow—to whom more presents had been made than to any of the others.

Some of the most decent of the men appeared really sorry at parting, and waded into the water to shake hands with me.

We got under weigh with a light air of wind from the N.E. at 25 minutes past 2. Our progress was very slow, there being frequent calms, so that, between pulling and sailing, we reached only to within five miles of Cape Hope at 4 A.M. of the 13th.

A large black whale and some white ones, with innumerable seals, were seen. Thermometer at +65°; but it became much colder after the wind came from sea. During the night we sailed among loose ice. As it was still calm we anchored at half-past 4 A.M. to wait for the other boat, which was some miles astern, to re-stow the cargo and cook breakfast. Thermometer at 5 A.M. +48°.

At half-past 6 we began pulling along shore. An hour afterwards a light breeze sprung up, but still ahead. The breeze becoming stronger, we hoisted sail and turned to windward, and would have made good progress had it kept steady, instead of which it followed or rather preceded the sun in his course westward, and thus headed us at every point we weathered. The flood-tide assisted us until 4 P.M., when we put ashore, as the ebb was too strong for us. Shot a young Arctic hare. There is a number of long narrow lakes near the point we stopped at, which is formed of grey and red granite and gneiss, and is about five miles from the S.E. point of Repulse Bay. Caught three species of marine insects with fins, which they use like wings: preserved specimens of them. Every appearance of rain this evening. Thermometer +65° at 8 P.M.

14th.—The wind shifted to the N.N.W. at half-past 9 last night, when we immediately got under weigh and sailed cautiously along shore, examining every bay and inlet when I supposed us near the northern outlet of Wager River, but not a trace of it was to be seen. If it exists, I think it not likely that it should have escaped our notice twice. The wind was for a few hours variable and squally; but it now shifted to N.E. by N. and blew hard. In crossing Wager River Bay, eight or ten miles from shore, there was a very heavy cross sea, which washed over our gunwales occasionally. On nearing the shore the run of the sea became more regular; but the wind increased so as to make it necessary to reef sails. The weather assuming a very threatening appearance, and the navigation being intricate and dangerous, we were forced to seek a harbour, which, after some difficulty, we found in a small bay at 8 P.M., having run from ninety to ninety-five miles, seventy-three of which were measured by Massey's patent log. Two white bears and many walruses were seen on a small island near Whale Island; but the weather was too stormy to permit us to pursue them.

It had been my intention to cross over to Southampton Island and trace that portion of the coast from Port Harding southward which had not yet been surveyed; but a stream of ice and the state of the weather prevented my doing so, nor did I think it an object of sufficient importance to detain the expedition a day or two for that sole purpose. Thermometer about +41° all day.

The male eider and king ducks appeared to have left this coast already, there being none but females seen. Our boat took the ground about half ebb—a fine bottom of sand and mud.

15th.—It blew a complete gale all night and during the greater part of this day. The sky, however, was sufficiently clear to allow me to obtain a meridian observation for latitude and variation. The former was found to be 64° 49' 06" N.—the latter 41° 27' W. Thermometer +46°.

The wind began to fall in the evening, and the tide having come in so as to float the boats, we started at 4 P.M. under reefed sails. The sea was still running high, but it was long and regular, and as there was every appearance of fine weather, I determined to sail all night, keeping a sharp lookout ahead for shoals, reefs, and islets. There was a heavy swell all night which broke with great violence on the reefs; and it being very dark, both boats were once or twice nearly filled by getting into shallow water before we were aware of it.

16th.—At half-past 5 this morning we were opposite Cape Fullerton, and at 6 Massey's log was examined, when it indicated a run of seventy-two miles. At 9 A.M. it fell calm. Thermometer +43°. An hour afterwards there was a light breeze from S.W., with which we turned to windward among numerous rocky islands.

At noon the latitude, 63° 56' 13" N., was observed, and shortly afterwards two Esquimaux were seen coming off in their kayaks, paddling at a great rate; but the breeze had now freshened, and it would have given them hard work to overtake us had we not shortened sail, and afterwards landed on an island, where we waited for them. Three more joined us there. They were very dirty, and far inferior in every respect to our friends of Repulse Bay. One of them was about five feet eight inches high, had a formidable beard and moustache, and was better looking than the others. After making them some presents we shoved off, and stood across the bay to the westward of Cape Fullerton. This bay is much deeper than it is laid down in the chart, and is crowded with islands.

It was near high water when we reached the main shore, and as we could make no progress against wind and tide, we put into a safe harbour. Nothing was to be seen for a mile or two inland but rocks, clothed in some spots with moss or grass. Deer were observed, and a young one shot by Nibitabo.

About an hour after our landing the wind shifted to W.N.W., and, as I was afraid of getting aground in our present berth, the boats were moved to a more open situation from which they could start at any time of tide.

The Esquimaux could tell us nothing about Churchill, none of them having visited that place either this or the previous summer. Thermometer at 9 P.M. +53°.

17th.—We were under weigh at 2 A.M., but the wind was both light and close, so that our progress was slow. Before the tide changed it came more from the southward; we were therefore obliged to anchor as soon as it began to ebb. The latitude of our harbour was 63° 47' 33" N. Var. 31° 8' W. The rocks, like those where we landed last night, were grey granite and gneiss. Thermometer at noon +60°. A large black whale was seen this morning.

At half-past 1 P.M. the tide began to flow, and at two we were under sail, the wind having gone round to the northward, so as to permit us to lie our course along shore. A succession of reefs lines the coast, which is itself very irregular in its outline, being indented with numberless inlets, some of them running many miles inland.

The tide began to ebb at 8 P.M., and as the wind had fallen and headed us, we ran in shore and cast anchor under the shelter of some rocks. It was just getting dark when a fresh breeze of fair wind sprung up. This was annoying enough. At 10 o'clock nine Esquimaux visited us, but staid only a short time, as we were to stop near their tents in the morning. Two of them said they would sleep on the rocks near us, with the intention of pointing out the deepest channel when we should resume our voyage.

18th.—We started at daylight this morning, but the fair wind, which had continued all night, soon failed us. Aided by the flood-tide, however, an hour's rowing brought us to the encampment of our last night's visitors, who welcomed us with much noise, and soon brought to the beach a number of furs and other articles for trade. They were very easy to deal with, apparently putting implicit confidence in our honesty; nor were they losers by this conduct. Ammunition was the article chiefly in demand, as they had two guns among the party. Files, knives, fire-steels, &c. were distributed among the men, and beads, needles, buttons, &c. among the women. One of the women was rather good-looking, but they were all much darker than the natives of Repulse Bay. They were well provided with food, as they had a large seal lying on the rocks, besides venison. It was still calm when we left them, but favoured by the ebb-tide we pulled out of the inlet, and shaped our course towards Chesterfield Inlet, which we crossed with the last of the flood. The day was beautiful—far too much so—and the few light airs of wind were all against us. We landed in a small cove on the south side of the inlet to pick up a deer that was shot from the boat. Four more deer were killed, but all in poor condition.

About two miles to the northward of the inlet I obtained a meridian observation of the sun in the natural horizon, which gave latitude 63° 32' 00" N. Thermometer at noon +65°, and in the evening +70°. The musquitoes were very numerous and troublesome. Numbers of turnstones (*Tringa interpres*) were seen.

19th.—There was a fine breeze again all last night, which died away at daylight. As soon as the flood-tide began to come in, we started with a light wind fair enough to allow us to lie our course along shore for a few miles. It again fell calm, when we took to the oars and landed on a point five miles to the southward of our last night's harbour, where we breakfasted at 9 A.M.

Dovekies in countless numbers were sitting on the stones, and swimming along the shore;[5] one or two pintailed and mallard ducks were seen on a lake a few hundred yards inland—the first we have seen since passing Nevill's Bay last year. Some dovekies' eggs were found with the birds formed in them.

Having obtained a meridian observation of the sun, which gave for the latitude 63° 17' 00" N., and variation 9° 21' W., we got under weigh and beat to windward with the last of the ebb, which here ran to the south. There was a fine breeze, but we made only about five miles southing, when at 6 P.M. the flood setting in strong against us, we put ashore for the night under the lee of the point. It was not easy to find a harbour, all the coast from Chesterfield Inlet being flat and stony, and lined with shoals. A young buck was shot, but it was in poor condition. Thermometer at noon +63°—at 8 P.M. +57°. Some of the copper came off our boat to-day and stopped her way before it was observed.

20th.—We were under weigh this morning by daylight, but the wind was right ahead and blowing fresh. Some more copper came off the boat, and she was evidently out of trim, as the Magnet went fast to windward of us. She had become leaky also, and therefore I determined to lay her aground as soon as the tide turned.

We had gained between six and seven miles, when, finding that we made but slow progress, I put on shore at the first place that offered shelter, a little before noon. Several deer were seen, and a large buck shot, which I was surprised to find very lean. At this season, near Repulse Bay they are in fine condition. Thermometer at noon +61°. At half-past 2 the wind changed to W.N.W., but it blew a gale before the tide flowed sufficiently to float us. We could do nothing but haul out into deeper water, to be ready by dawn next morning.

Some pintails, mallards, and Hutchins and laughing geese were seen here; also a brood of well-grown young king-ducks in a small lake at some distance from the sea, with which it had no connection.

Just as our boats floated, the wind became more moderate; and as we had still an hour and a half of daylight, we sailed along the coast for 4½ miles, being forced to keep some miles from shore to avoid shoals. Soon after sunset we ran into a bay for shelter during the night. In doing so we grazed some ridges of stones, but found good anchorage in four fathoms water. Thermometer +47°.

21st.—Thermometer +44°. There was a strong breeze with heavy squalls from the north all night. On starting at daylight and making for the only outlet that appeared, we found it too shallow, and so were forced to wait the flow of the tide. The wind was W. by N., but gradually shifted round against us and became very light. We managed, however, to reach an island near the north point of Rankin's Inlet.

Although there was a fine breeze, it being right ahead, nothing was to be gained against the ebb tide.

We found many old signs of Esquimaux visits to the island. Among other articles picked up were an ivory snow-knife, a drill for producing fire, and an iron drill; also some vertebræ of a whale measuring ten inches in diameter. There were numerous graves of Esquimaux here, with spears, lances, &c. deposited beside them. Most of these articles were old and much corroded with rust, but a very excellent seal-spear head had been placed there this spring. Thermometer at noon +52°; 8 P.M. +47°. Temperature of water +41°.

22d.—Thermometer +42°. At a little before 5 this morning the wind shifted to S.S.E. We set out to cross Rankin's Inlet, although we could not lie our course, and after five hours' sailing reached an island near the south shore, where we landed, as the breeze had increased to a gale and gone more to the southward, with a heavy sea, which washed over us occasionally. We here picked up some specimens of copper ore, but the ore did not appear to be abundant.

The aurora was very bright last night. It appeared first to the S.S.E., moved rapidly northward, spreading all over the sky, and finally disappearing in the north. This agrees with what Wrangel asserts, "that the aurora is affected by the wind in the same way as clouds are." Heavy rain and a strong gale from noon until 8 P.M. Temperature of water +42°; air +43°.

23rd.—The wind was right ahead but light this morning. We got under weigh and beat to windward some miles, alternately sailing and pulling until we reached the north point of Corbett's Inlet. We were here visited by eighteen Esquimaux in their kayaks. All the news they could give us was that one of Ouligbuck's sons had passed the winter near this place, and that he had walked to Churchill in the winter, where all were then well. A brisk trade was soon opened; the articles in greatest request being powder and ball. Some fox and wolf skins were received, but before they had brought out the half of their stock, the wind changed from S.W. to N.W. by W. and blew a gale, which soon raised a sea that washed over the canoes alongside. Being anxious to take advantage of the fair wind to cross Corbett's Inlet before dark, after making our friends presents of various articles, we set sail and ran across the inlet, encountering a heavy sea caused by a swell from the south meeting the waves raised by the present gale. We were three hours crossing to the south point of the inlet, off which lie some dangerous reefs five or six miles from land. The wind was very close as we turned the point, and after gaining six miles further, we were forced to make a number of tacks before getting into a harbour, which proved to be an excellent one, land-locked on all sides. Little soil was to be seen on the rocks, which were of granite. We had shipped a good deal of water, and it was past 9 P.M. when we got under shelter. Thermometer +45°. Hundreds of grey phalaropes were seen, supposed to be Phalaropus fulicarius.

24th.—It blew so hard this morning that we could not start until 8 o'clock. The wind after that moderated gradually, and latterly fell calm. By rowing we arrived at the S.E. end of the island[6] near Whale Cove, where we were visited by a party of natives, who brought off some furs and boots for trade. A breeze from S.S.E. sprung up about 1 o'clock, with which we turned to windward through a narrow channel between a small island and the main. When we reached the open sea the wind was too much ahead for us to advance against the ebb tide, and as a convenient harbour offered itself, we anchored for the night. Our latitude at noon was 62° 13' 19", after which we advanced about four miles to the southward. Ouligbuck told us that, when a little boy about seven years old, he visited this place with his parents, and went out to Sea-Horse Island on the ice to hunt the animals from which it takes its name. Three large black whales were seen to-day. Thermometer +46°, +53°, and +42°.

I was much pleased to observe that the nearer we approached to Churchill, the more confidence the Esquimaux placed in us. They fixed no price for their goods, but threw them on board the boat, and left it to me to pay them what I pleased. This confidential mode of dealing, which is not in keeping with the habits of the Esquimaux tribes, at least shows that they are satisfied with the treatment they receive at Churchill. To the Hudson's Bay Company, indeed, they have much reason to be grateful for having, by their influence, at last created a friendly feeling between them and the Chipewyans, with whom they used to be at constant and deadly enmity.

25th.—There was heavy rain all last night, which continued until between 9 and 10 o'clock this morning. We then got under weigh with the first of the flood, but it fell calm. We rowed for fourteen or fifteen miles, the rain pouring all the time. A fine breeze from N. by E. sprung up at 4 P.M., before which we ran direct for the passage between Sir Bibye's Islands, but finding the water become very shallow, and learning from Ouligbuck that there was not water enough for boats except at full tide, we kept outside the islands altogether. We reached the main land a little after sunset at the south point of Nevill's Bay, and ran for shelter into a small inlet separated on the south by a narrow point from a deep river, to which the Esquimaux resort to catch salmon. Thermometer +37° and +41°. As the moon was full, I at first intended running on all night, but the threatening look of the weather deterred me.

26th.—Last night, about an hour after casting anchor, the moon became overcast, and it blew a perfect gale. On landing this morning we found a quantity of wood, a large sledge 30 feet long, and some slender pieces of wood fastened together to the length of 40 feet. There were two of these poles, which are used by the natives for spearing small seals. It is said that, in Davis' Straits, the Esquimaux use poles of the same kind for spearing whales.

As the bay in which we were lying was not very safe should the wind change, we got under weigh and turned into the mouth of the river under close-reefed sails. The boats shipped much water, particularly the Magnet, keeping a man constantly baling. We at last got under the lee of a point where there was a sandy bottom, but not water enough to float the boats at low tide. The river is about a mile broad, and deep enough in the middle for a vessel drawing 12 or 14 feet water.

We saw a number of whalebone snares set along the edges of the lakes for geese, large flocks of which were feeding about, but very shy. There was a storm from N.N.W. all the afternoon with heavy rain. Thermometer +36°.

27th.—It felt very cold this morning; the thermometer was at the freezing point, and there was some snow. The storm had continued all night with increasing strength, but towards day-light the weather became more moderate, so that about 9 o'clock we were able to start under reefed sails. The breeze gradually died away and went round to the S.W., and it finally became calm. Heavy rain and sleet began to fall; the wind veered round to the S.E., so that we could lie our course, and make good progress with the flood.

At 6 P.M. we reached a bay a few miles north of Knapp's Bay, which I had not noticed on our outward voyage, and which is not laid down on the charts. It is about ten miles wide and eight deep; the water in it is very shallow, no where exceeding ten feet; and as it was within an hour or two of high water, the greater part of it must be dry when the tide is out.

Numbers of Brent geese were feeding in all directions on a marine plant (*zostera marina*, Linn.) which grows here in great abundance.

We anchored under the lee of an islet in Knapp's Bay, a very small portion of which was visible at high water. Thermometer +38°.

28th.—We were under weigh at day-light this morning, with a strong breeze of north-west wind, which made us close-reef our sails. There was a heavy sea in Knapp's Bay. At 8 A.M. we passed to the westward of the island, under which we found shelter during the gale of the 8th of July last. The wind was cold, with occasional showers of rain. Great numbers of geese were seen passing to the southward. In the evening the wind became more moderate and finally calm. Our water-kegs being empty, I ran inshore a little before sunset, and entered Egg River, in which we found a safe harbour. This river discharges a considerable body of water into the sea by five mouths, separated by four islets. There is no island lying opposite to its mouth, as represented in the charts. Thermometer from +35° to +40°.

29th.—The boat lay afloat all the night, which was fine but dark. There was not a breath of wind until 7 o'clock. An hour after starting, a moderate breeze sprung up from W. by N., but soon became light and variable, and at last it fell calm a short time before sunset, when, having gained about 40 miles, we pulled into a small bay, which afforded us good shelter. The day was fine throughout, with occasional light showers of rain. Thermometer from +45° to +52°.

The sky was too much overcast for me to obtain any observation, but it appears to me that Egg River is laid down in the charts about 12 miles too far to the southward, and Egg Island is 12 miles south of the river instead of being near its mouth, as there represented.

30th.—We had 13 feet water last night when the tide was in, but it was not until the flood had made two hours that we floated. The night was as fine as the last and calm. There was a light air of west wind when we got under weigh, with which and the flood-tide we slipped alongshore pretty fast. In an hour or two the wind began to fly about from all points, with calms between, so that even with the help of our oars we only made 22 miles, and not being able to reach Seal River, we ran into a small bay—the only spot that appeared clear of stones for some miles—about 12 miles north of it. Here abundance of drift wood was found, with which the men lighted fires sufficiently large for the coldest winter night. The evening was very warm, and the musquitoes were troublesome. The country inland is well wooded. Great numbers of mallard, teal, pintails, and long-tailed ducks were seen, but only two or three were shot.

31st.—Left our harbour as soon as the tide permitted, which was at 7 A.M. A light but fair breeze from N. by W. gradually increased, so that we made a fine run across Button's Bay, which is as full of rocks and shoals as represented in the charts, and entering Churchill River a few minutes after 1 P.M., landed in a small cove a few hundred yards above the Old Fort.

On visiting the Company's establishment, I found that Mr. Sinclair was absent at York Factory; but I was very kindly received by Mrs. Sinclair, and liberally supplied with everything we required for the continuation of our voyage. As we had carried away our bowsprit, Turner was set to make a new one.

I received many letters from much valued friends, and after remaining for a few hours, returned to the boats at 9 P.M. in order to be prepared for starting early in the morning, should wind and weather prove favourable. The stock of provisions on hand was eight bags of pemmican and four cwt. of flour. We left Ouligbuck and his son at Churchill.

3rd September.—For the last two days the wind had been fair, but blowing a gale, with such a heavy sea that we could not proceed. The weather was so cloudy that I could obtain no observations; I therefore employed most of my time in shooting Esquimaux curlews, which were so abundant near the Old Fort that I bagged seven brace in a few hours.

This morning the wind shifted more to the westward, and becoming more moderate, we got under weigh at 9 A.M. There was still a heavy swell outside and at the entrance of our little harbour. Whilst coming out in the dawn of the morning three seas came rolling in one after the other, and broke completely over the bows of the boat, washing her from stem to stern. I thought she would have filled, but we got into deep water before any more seas caught her. The Magnet was even more roughly handled in following us, having shipped much water and struck heavily on the rocks—fortunately without damage. The wind died away, and during the morning shifted to south. We, however, reached Cape Churchill, and at 8 P.M. cast anchor under its lee, exactly opposite an old stranded boat.

4th.—We had a breeze from S.W. by S. to-day, which enabled us to get along the coast sixteen or eighteen miles during the flood. It blew so hard in the afternoon that we required to double-reef our sails. The weather was very warm, the thermometer being as high as +60° in the shade. A Canada nuthatch (*sitta Canadensis*) flew on board to-day, and was very nearly caught. There were a good many ducks and geese near the place where we landed to get fresh water. Between thirty and forty of the former and two of the latter were shot. The boats were allowed to take the ground, after two hours' ebb, on a fine shingle beach, on which a considerable surf was breaking.

5th.—It was calm all night. At 3 this morning the boat floated, and we pushed out a short distance from shore to be ready for the first fair wind. At half-past seven a light air sprung up from N.E., but did not increase till past noon, when there was a fine breeze. A meridian observation of the sun gave latitude 58° 26' 14" N. At 5 P.M. we were opposite the mouth of Broad River, latitude 58° 7' 0" N. Thermometer at noon +56°.

6th.—We were under weigh this morning a little before daylight with the wind from N.E. The weather was so thick that we could not see more than a hundred yards ahead. We, however, ran on by soundings until I thought we were near North River, and then kept inshore until we got sight of land, which proved to be close to Nelson River, across which we stood, directing our course by compass, and coming in directly opposite the beacon. We arrived at York Factory between 9 and 10 o'clock P.M. and were warmly welcomed by our friends, who had not expected to see us until next summer.

In justice to the men under me, let me here express my thanks for their continued good conduct under circumstances sometimes sufficiently trying;—in fact, a better set of fellows it would be difficult to find anywhere.

As to their appearance when we arrived at York Factory, I may adopt the words of Corporal McLaren in charge of the Sappers and Miners who are to accompany Sir John Richardson,—"By George, I never saw such a set of men."

ENDNOTES

The following notes appeared as footnotes in the original publication of *Narrative of an expedition to the shores of the Arctic Sea, in 1846 and 1847.*

[1] The male and female of the northern diver (*colymbus glacialis*) resemble one another so much that it is very difficult to distinguish the one from the other. The immature bird has often been described by ornithologists as the female.

[2] These birds breed in great numbers among the rocks in Orkney, and are much attached to their young. By chasing the latter in a boat they become so fatigued as to be easily caught. When one of them is taken into the boat the parent bird approaches within a few feet, dives under and around the boat in all directions, and whenever it comes up to the surface utters a peculiarly melancholy note, at the same time turning its head in a listening attitude as if expecting to hear an answer from the prisoner. The anxiety of the mother has always the desired effect, and it is pleasing to observe the joy with which she swims away with her recovered young one, nestling it under her wing and never permitting it to stray a foot from her.

[3] An excellent plan of shooting these birds, and one that I have often successfully practised, is to roll up a bit of fur or cloth about the shape and size of a mouse, and drag it after you with a line twenty yards long. The owl will soon perceive the decoy, although half-a-mile distant; and after moving his head backwards and forwards as if to make sure of his object, he takes wing, and making a short sweep in the rear of his intended prey, pounces upon and seizes it in his claws, affording the sportsman a fine opportunity of knocking him down. I have sometimes missed my aim, leaving the owl to fly away with the false mouse (which the sudden jerk had torn from the line) in his claws. The Indians, taking advantage of this bird's propensity to alight on elevated spots, set up pieces of wood in the plains or marshes with a trap fastened to the top. In this way I have known as many as fifty killed in the early part of winter by one Indian. The owl is very daring when hungry. I remember seeing one of these powerful birds fix its claws in a lapdog when a few yards distant from the owner, and only let go his gripe after a gun was fired. The poor little dog died of its wounds in a few days.

4 Receipt.—Seven lbs. flour, 1 oz. carbonate soda, ¾ oz. citric acid, ¾ oz. common salt, water (cold), about ½ gallon. The salt, soda, and acid being finely powdered and dry, are to be well mixed together; this mixture being well wrought up with the dry flour, the water is to be added in 2 or 3 parts and mingled with the flour as quickly as possible; the dough being put into pans is immediately to be placed in the oven.

5 The dovekie, or black guillemot (*Uria grylle*), breeds in great numbers in the Orkney islands. I believe ornithologists are mistaken in supposing that this bird becomes white or rather grey during the winter. It is only the young birds that are so, the old ones are seen in winter without any change in the colour of their summer plumage.

6 This place is laid down on the chart as an island, but is a peninsula according to the account we received from the Esquimaux.

APPENDIX

John Rae kept exhaustive records of everywhere he went and everything he saw. Beyond the narrative are detailed lists of plants and animals, included below, and beginning on page 194, a number of tables recording the "dip of the needle and force of magnetic attractions at various stations" as well as temperature readings from September 1846–August 1847.

LIST OF MAMMALIA,
Collected during Mr. Rae's Expedition, with Observations by J.E. Gray, Esq., F.R.S. &c.

1. *Mus Musculus.* Linn. York Factory. Probably introduced from Europe.

2. *Arctomys Parryi.* Richardson, Faun. Bor. Amer. p. 158, tab. 10.

3. *Lepus Glacialis.* Leach. Richardson, Faun. Bor. Amer. 221.
   Myodes.—The specimens brought by the expedition have enabled me to make some corrections in the characters assigned to these species. I may observe that the large size or peculiar form of the claws which has been regarded as a character of the species, appears to be peculiar to one sex—probably the males.
   1. *The upper cutting teeth narrow, smooth without any longitudinal groove. Thumb with a compressed curved acute claw.* (Lemnus).
   Myodes, Lemnus Pallas. Glires 77 of Sweden.
   Myodes Helvolus. Richardson, Faun. Bor. Amer. p. 128, belong to this section. All the museum specimens of these species have small, simple, curved, acute claws.

4. *Myodes Hudsonius.* Richardson, Faun. Bor. Amer. 132.
   Grey, black washed beneath white, sides reddish, sides of the neck red, nose with a central black streak, claws of male (?) very large, compressed, equal, broad to the end, and notched; of female small, acute. In winter with very long black white-tipped hairs. Mr. Rae brought home two males, one in winter and one in change fur, and two females in summer fur.

5. *Myodes Greenlandicus.*
Reddish-grey, brown, black varied, back with a longitudinal black streak, beneath grey brown, chest, nape, and sides ruffous. Front claw of males (?) compressed, curved, the under surface (especially of the middle one) with a broad, round, expanded tubercle. I have not seen this species showing any change in its winter fur.
2. Upper cutting teeth broader, with a central longitudinal groove. The claw of the front thumb strapshaped, truncated, and notched at the tip.

6. *Myodes Helvolus.* Richardson, Faun. Bor. Amer. 128. (female ?)
Fur very long, black, grey-brown; black grizzled, hinder part of the body reddish, beneath grey, sides yellowish. Claws of the fore feet (of the males ?) large, thick, rounded, curved, bluntly truncated at the tip; of the female compressed, curved, acute.

7. *Myodes Trimuconatus.* Richardson, Faun. Bor. Amer. 130.
Bright red brown, head blackish-grey, sides and beneath pale ruffous, chin white, claws moderate, compressed. This species is best distinguished from the former by its larger size and the great brightness of the colour, and the fur being much shorter and less fluffly.

LIST OF THE SPECIES OF BIRDS,

Collected by Mr. Rae during his late Expedition, named according to the "Fauna Boreali-Americana," by G.R. Gray, Esq., F.L.S.

Falconidæ.
  Aquila (Pandion) halisæeta.
  Falco peregrinus.
  "   islandicus.
  Accipiter (Astur) palumbarius.
  Buteo lagopus.
  "   (Circus) cyaneus.

Strigidæ.
  Strix brachyota.
  "   funerea.
  "   Tengmalmi.

Janiadæ.
  Tyrannula pusilla.

Merulidæ.
  Merula solitaria.

Sylviadæ.
  Sylvicola æstiva.
  "   coronata.
  Sylvicola striata.
  "   (Vermivora) rubricapilla.
  "   "   peregrina.
  Seiurius aquaticus.
  Anthus aquaticus.

Fringillidæ.
  Alauda cornuta.
  Emberiza (Plectrophanes) nivalis.
  "   "   lapponica.
  "   "   picta.
  "   canadensis
  "   (Zonotrichia) leucophrys.
  "   "   pennsylvanica.
  "   "   iliaca.
  Fringilla hyemalis.
  Pyrrhula (Corythus) enucleator.
  Logia leucoptera.
  Linaria minor.

Sturnidæ.
  Quiscalus versicolor.
  Scolecophagus ferrugineus.

Corvidæ.
  Garrulus canadensis.

Picidæ.
  Picus (Apternus) tridactylus.
  Colaptes auratus.

Rasores.
  Tetrao canadensis.
  "   (Lagopus) mutus.
  "   "   saliceti.
  "   (Centrocercus) phasianellus.

Grallatores.
  Calidris arenaria.
  Charadrius semipalmata.
  Vanellus melanogaster.
  Strepsilas interpres.
  Tringa Douglassii.
  "   maritima.
  "   alpina.
  "   Schinzii.
  "   pusilla.
  "   cinerea.
  Totanus flavipes.
  "   macularius.
  Limosa hudsonica.
  Scolopax Wilsoni.
  Phalaropus hyperboreus
  "   fulicarius.

Natatores.
  Podiceps cornutus.
  Larus argentatoides.
  Lestris pomarina.
  "   parasitica.
  "   Richardsoni.

Anas (Boschas) crecca, var.
" " discors.
Somateria spectabilis.
" mollissima.
Oidemia perspicillata.
" americana.
Harelda glacialis.
Mergus serrator.
Anser albifrons.
" hyperboreus.
" Hutchinsii.

" bernicla.
Colymbus arcticus.
" septentrionalis.
Myiodioctes pusilla.
Regulus calendula.
Sitta canadensis.
Linaria borealis.
Tringa rufescens.
" pectoralis.
Totanus solitarius.

FISHES,
Collected during Mr. Rae's Expedition. By J.E. Gray, Esq., F.R.S.

Gadidæ.
Lota Maculosus. Richardson, Faun. Bor. Amer. iii. 248. Male and female.

Esocidæ.
Esox. Lucius. Richardson, Faun. Bor. Amer. iii. 124. Female.

Cyprinidæ.
Catastomus Forsterianus? Richardson, Faun. Bor. Amer. iii. 116. Female. Lakes near York Factory. The "Red Sucker."
Catastomus Hudsonius. Richardson, Faun. Bor. Amer. iii. 112. River near York Factory. "The Grey Sucker."

Salmonidæ.
Salmo. Salar? Richardson, Faun. Bor. Amer. 145. Repulse Bay.
Salmo Hoodii. Richardson, Faun. Bor. Amer. iii. 173, t. 82, f. 2, t. 83, f. 2, t. 87, f. 1. Male and female. Lakes near York Factory.
Salmo Coregonus Albus. Richardson, Faun. Bor. Amer. 195. t. 89, f. 2. a. b. Male. The Attihawmeg. Lower jaw shortest; ridge behind the eye becoming close to the orbit beneath the eye.
Salmo (Coregonus) Tullibee. Richardson, Faun. Bor. Amer. 201. Lakes near York Factory. "The Tullibee." Lower jaw shortest, ridge behind continued distant from the orbit and produced towards the nostrils.
Salmo Coregonus Harengus? Richardson, Faun. Bor. Amer. 210. t. 90, f. 2, a. b. Lower jaw longest, ridge behind the eyes becoming rather nearer to, but distinct from, the orbit beneath. River near York Factory.

PLANTS,

Named by Sir W.J. Hooker, K.H., D.C.L., F.R.A. & L.S. &c. &c. &c.

Plants collected on the Coast between York Factory and Churchill, and in the neighbourhood of Churchill.

DICOTYLEDONES.

Ranunculaceæ, *Juss.*
1. Anemone *Richardsoni*, Hook. Fl. Bor. Am. i. 6, Tab. 4, A.
2. Ranunculus *Lapponicus, L.*—Hook. Fl. Bor. Am. i. p. 16.

Cruciferæ, *Juss.*
3. Nasturtium *palustre*, De Cand.—Hook. Fl. Bor. Am. i. p. 39.
4. Arabis *petræa*, Lam.—Hook. Fl. Bor. Am. i. p. 42.
5. Cardamine *pratensis*, L.—Hook. Fl. Bor. Am. i. p. 45.
6. Draba *hirta*, L.—Hook. Fl. Bor. Am. i. p. 52.
7. Draba *alpina*, L.—Hook. Fl. Bor. Am. i, p. 50.

Caryophylleæ, *Juss.*
8. Stellaria *Edwardsii*, Br.—Hook. Fl. Bor. Am. i. p. 96, Tab. 31.
9. Cerastium *alpinum*, L.—Hook. Fl. Bor. Am. i. p. 104.
10. Silene *acaulis*, L.—Hook. Fl. Bor. Am. i. p. 87.
11. Arenaria *peploides*, L.—Hook. Fl. Bor. Am. i. p. 102.

Leguminosæ, *Juss.*
12. Phaca *astragalina*, De Cand.—Hook. Fl. Bor. Am. i. p. 145.
13. Oxytropis *campestris*, De Cand.—Hook. Fl. Bor. Am. i. p. 147.
14. Oxytropis *deflexa*, De Cand.—Hook, Fl. Bor. Am. i. p. 148.
15. Hedysarum *Mackenzii*, Rich.—Hook. Fl. Bor. Am. i. p. 155.

Rosaceæ, *Juss.*
16. Dryas *integrifolia*, Vahl.—Hook. Ex. Fl. Tab. 200, Fl. Bor. Am. i. p. 174.
17. Rubus *acaulis*, Mich.—Hook. Fl. Bor. Am. i. p. 182.
18. Potentilla *anserina*, L.—Hook. Fl. Bor. Am. i. p. 189.
19. Potentilla *pulchella*, Br.—Hook. Fl. Bor. Am. i. p. 191.
20. Potentilla *nivea*, L.—Hook. Fl. Bor. Am. i. p. 195.

Onagrarieæ, *Juss.*
21. Epilobium *latifolium*, L.—Hook. Fl. Bor. Am. i. p. 205.

Saxifrageæ, *Juss.*
22. Saxifraga *oppositifolia*, L.—Hook. Fl. Bor. Am. i. p. 242.
23. Saxifraga *cæspitosa*, L.—Hook. Fl. Bor. Am. i. p. 244.
24. Saxifraga *Hirculus*, L.—Hook. Fl. Bor. Am. i. p. 252.
25. Saxifraga *tricuspidata* L.—Hook. Fl. Bor. Am. i. p. 254.

Compositæ, *Juss.*
26. Nardosmia *corymbosa*, Hook. Fl. Bor. Am. i. p. 307 (Tussilago corymbosa, Br.)
27. Achillæa *millefolium*, L.—Hook. Fl. Bor. Am. i. p. 318.
28. Chrysanthemum *arcticum*, L.—Hook. Fl. Bor. Am. i. p. 319.
29. Pyrethrum *inodorum*, Sm.—Hook. Fl. Bor. Am. i. p. 320.
30. Senecio *aureus*, L.—Hook. Fl. Bor. Am. i. p. 333. var. nanus.
31. Arnica *montana*, L.—B. *angustifolia*, Hook. Fl. Bor. Am. i. p. 330.

Campanulaceæ, *Juss.*
32. Campanula *uniflora*, L.—Hook. Fl. Bor. Am. ii. p. 29.

Ericeæ, *L.*
33. Ledum *palustre*, L.—Hook. Fl. Bor. Am. ii. p. 44.—var. a. *angustifolium*; and var. B. *latifolium*.
34. Azalea *procumbens*, L.—Hook. Fl. Bor. Am. ii. p. 44.
35. Rhododendron *Lapponicum*, Wahl.—Hook. Fl. Bor. Am. ii. p. 43.
36. Vaccinium *Vitis Idæa*, L.—Hook. Fl. Bor. Am. ii. p. 34.

Monotropeæ, *Nutt.*
37. Pyrola *rotundifolia*, L.—Hook. Fl. Bor. Am. ii. p. 46.

Boragineæ, *Juss.*
38. Lithospermum *maritimum* Lehm.—Hook. Fl. Bor. Am. ii. p. 86.

Scrophularineæ, *Juss.*
39. Castilleja *pallida*, Benth.—Hook. Fl. Bor. Am. ii. p. 105.
40. Bartsia *alpina*, L.—Hook. Fl. Bor. Am. ii. p. 106.
41. Pedicularis *Wlassoviana*, Stev.—Hook. Fl. Bor. Am. ii. p. 107.
42. Pedicularis *Lapponica*, L.—Hook. Fl. Am. ii. p. 108.
43. Pedicularis *Sudetica*, Willd.—Hook. Fl. Bor. Am. ii. p. 109.
44. Pedicularis *flammea*, L.—Hook. Fl. Bor. Am. ii. p. 110.
45. Pedicularis *euphrasioides*, Stev.—Hook. Fl. Bor. Am. ii. p. 108.

Primulaceæ, *Juss.*
46. Androsace *septentrionalis*, L.—Hook. Fl. Bor. Am. ii. p. 119.
47. Primula *Hornemanniana*, Lehm.—Hook. Fl. Bor. Am. ii, p. 120.

Polygoneæ, *Juss.*
48. Polygonum *viviparum*, L.—Hook. Fl. Bor. Am. ii. p. 130.

Amentaceæ, *Juss.*
49. Salix *Richardsoni* Hook. Fl. Bor. Am. ii. p. 147, Tab. 182.
50. Salix *vestita*, Ph.—Hook. Fl. Bor. Am. ii. p. 152.
51. Salix *Arctica*, Br.—Hook. Fl. Bor. Am. ii. p. 152.
52. Betula *glandulosa*, Mx.—Hook. Fl. Bor. Am. ii. p. 156.
53. Betula *nana*, L.—Hook. Fl. Bor. Am. ii. p. 156.

MONOCOTYLEDONES.

Melanthaceæ, *Br.*
54. Tofieldia *palustris*, Huds.—Hook. Fl. Bor. Am. ii. p. 179.

Orchideæ, *Juss.*
55. Platanthera *obtusata*, Lindl.—Hook. Fl. Bor. Am. ii. p. 196, Tab. 199.
56. Platanthera *rotundifolia*, Lindl.—Hook. Fl. Bor. Am. ii. 200, Tab. 201.

Cyperaceæ, *Juss.*
57. Carex *dioica*, L.—Hook. Fl. Bor. Am. ii. p. 208.
58. Carex *fuliginosa*, Sternb. and Hoppe.—Hook. Fl. Bor. Am. ii. p. 224.
59. Eriophorum *capitatum*, Host,—Hook. Fl. Bor. Am. ii. p. 231.
60. Eriophorum *polystachyon*, L.—Hook. Fl. Bor. Am. ii. p. 231.

Collected between Churchill and Repulse Bay.

DICOTYLEDONES.
Ranunculaceæ, *Juss.*
1. Ranunculus *affinis*, Br.—Hook. Fl. Bor. Am. i. p. 12, Tab. 6 A.

Papaveraceæ, *Juss.*
2. Papaver *nudicaule*, L.—Hook. Fl. Bor. Am. i. p. 34.
3. Arabis *petræa*, Lam.—Hook. Fl. Bor. Am. i. p. 42.
4. Cardamine *pratensis*, L.—Hook. Fl. Bor. Am. i. p. 45.
5. Draba *alpina*, L.—Hook. Fl. Bor. Am. i. p. 50.
6. Eutrema *Edwardsii*, Br.—Hook. Fl. Bor. Am. i. p. 67.

Caryophylleæ, *Juss.*
7. Silene *acaulis*, L.—Hook. Fl. Bor. Am. i. p. 89.
8. Lychnis *apetala*, L.—Hook. Fl. Bor. Am. i. p. 94.
9. Stellaria *Edwardsii*, Br.—Hook. Fl. Bor. Am. i. p. 96. Tab. 31.
10. Cerastium *alpinum*, L.—Hook. Fl. Bor. Am. i. p. 104.

Leguminosæ, *Juss.*
11. Oxytropis *campestris*, De Cand.—Hook. Fl. Bor. Am. i. p. 146.
12. Oxytropis *Uralensis*, De Cand.—Hook. Fl. Bor. Am. i. p. 145.
13. Phaca *astragalina*, De Cand.—Hook. Fl. Bor. Am. i. p. 145.

Rosaceæ, *Juss.*
14. Dryas *integrifolia*, Vahl.—Hook. Fl. Bor. Am. i. p. 174.
15. Rubus *Chamæmorus*, L.—Hook. Fl. Bor. Am. i. p. 183.
16. Potentilla *nana*, Lehm.—Hook Fl. Bor. Am. i. p. 194.

Onagrarieæ, *Juss.*
17. Epilobium *latifolium*, L.—Hook. Fl. Bor. Am. i. p. 205.

Saxifrageæ, *Juss.*
18. Saxifraga *oppositifolia*, L.—Hook. Fl. Bor. Am. i. p. 242.
19. Saxifraga *cæspitosa*, L.—Hook. Fl. Bor. Am. i. p. 246.
20. Saxifraga *cernua* L.—Hook. Fl. Bor. Am. i. p. 246.
21. Saxifraga *rivularis* L.—Hook. Fl. Bor. Am. i. p. 246.
22. Saxifraga *Hirculus*, L.—Hook. Fl. Bor. Am. i p. 252. and var. bi-triflora.
23. Saxifraga *tricuspidata*, L.—Hook. Fl. Bor. Am. i. p. 253.

Compositæ, *Juss.*
24. Leontodon *Taraxacum*, L.—Hook, Fl. Bor. Am. i. p. 296.
25. Chrysanthemum *integrifolium*, Rich.—Hook. Fl. Bor. Am. i. p. 319, Tab. 109.
26. Erigeron *uniflorus*, L.—Hook. Fl. Bor. Am. ii. p. 17.

Campanulaceæ, *Juss.*
27. Campanula *uniflora*, L.—Hook. Fl. Bor. Am. ii. p. 29.

Ericeæ, *Juss.*
28. Andromeda *tetragona*, L.—Hook. Fl. Bor. Am. ii. p. 38.
29. Ledum *palustre*, L.—Hook. Fl. Bor. Am. ii. p. 44. var. *angustifolium.*

Diapensiaceæ, *Lindl.*
30. Diapensia *Lapponica*, L.—Hook. Fl. Bor. Am. ii. p. 76.

Boragineæ, *Juss.*
31. Lithospermum *maritimum*, Lehm.—Hook. Fl. Bor. Am. ii. p. 36.

Scrophularineæ, *Juss.*
32. Pedicularis *hirsuta*, L.—Hook. Fl. Bor. Am. ii. p. 109.
33. Pedicularis *Langsdorffii*, Fisch.—Hook. Fl. Bor. Am. ii. p. 109.

Plumbagineæ , *Juss.*
34. Statice *Armeria*, L.—Hook. Fl. Bor. Am. ii. p. 123.

Ameniaceæ, *Juss.*
35. Salix *Myrsinites*, L.—Hook. Fl. Bor. Am.ii. p. 151.
36. Salix *Arctica*, Br.—Hook. Fl. Bor. Am. ii. p. 152.

MONOCOTYLEDONES.
Junceæ, *Juss.*
37. Luzula *hyperborea*, Br.—Hook. Fl. Bor. Am. ii. p. 188.

Cyperaceæ, *Juss.*
38. Carex *membranacea*, Hook. Fl. Bor. Am. ii. p. 220.
39. Eriophorum *polystachyon*, L.—Hook. Fl. Bor. Am. ii. p. 231.

Gramineæ, *Juss.*
40. Alopecurus *alpinus*, L.—Hook. Fl. Bor. Am. ii. p. 234.
41. Hierochloe *alpina*, Ræm. et Sch.—Hook. Fl. Bor. Am. ii. p. 234.
42. Colpodium *latifolium*, Br.—Hook. Fl. Bor. Am. ii. p. 238.
43. Poa *Arctica*, Br.—Hook. Fl. Bor. Am. ii. p. 246.
44. Festuca *brevifolia*, Br.—Hook. Fl. Bor. Am. ii. p. 250.
45. Elymus *arenarius*, L.—Hook. Fl. Bor. Am. ii. p. 255.

Plants collected between Repulse Bay and Cape Lady Pelly.

DICOTYLEDONES.
Ranunculaceæ, *Juss.*
1. Ranunculus *Lapponicus*, L.—Hook. Fl. Bor. Am. i. p. 16.

Papaveraceæ, *Juss.*
2. Papaver *nudicaule*, L.—Hook. Fl. Bor. Am. i. p. 34.

Cruciferæ, *Juss.*
3. Cardamine *pratensis*, L.—Hook. Fl. Bor. Am. i. p. 44.
4. Draba *alpina*, L.—Hook. Fl. Bor. Am. i. p. 50.
5. Draba *stellata*, Jacq.—Hook. Fl. Bor. Am. i. p. 53.

Caryophylleæ, *Juss.*
6. Stellaria *humifusa*, Rottb.—Hook. Fl. Bor. Am. i. p. 97.
7. Cerastium *alpinum*, L.—Hook. Fl. Bor. Am. i. p. 104.

Leguminosæ, *Juss.*
8. Oxytropis *Uralensis*, De Cand.—Hook. Fl. Bor. Am. i. p. 145.
9. Oxytropis *campestris*, De Cand.—Hook. Fl. Bor. Am. i. p. 147.

Rosaceæ, *Juss.*
10. Dryas *integrifolia*, Vahl,—Hook. Fl. Bor. Am. i. p. 174.
11. Potentilla *nana*, Lehm.—Hook. Fl. Bor. Am. i. p. 190.

Onagrarieæ, *Juss.*
12. Epilobium *latifolium*, L.—Hook. Fl. Bor. Am. i. p. 204.

Saxifrageæ, *Juss.*
13. Saxifraga *oppositifolia*, L.—Hook. Fl. Bor. Am. i. p. 242.
14. Saxifraga *cernua*, L.—Hook. Fl. Bor. Am. i. p. 245.

15. Saxifraga *rivularis*, L.—Hook. Fl. Bor. Am. i. p. 246.
16. Saxifraga *nivalis*, L.—Hook. Fl. Bor. Am. i. p. 248.
17. Saxifraga *foliolosa*, Br.—Hook. Fl. Bor. Am. i. p. 251.
18. Saxifraga *Hirculus*, L.—Hook. Fl. Bor. Am. i. p. 252.

Compositæ, *Juss.*
19. Leontodon *Taraxacum*, L.—Hook. Fl. Bor. Am. i. p. 296.
20. Pyrethrum *inodorum*, Sm.—Hook. Fl. Bor. Am. i. p. 320.
21. Arnica *montana*, L.—B. *angustifolia*, Hook, Fl. Bor. Am. i. p. 330.
22. Erigeron *uniflorus*, L.—Hook. Fl. Bor. Am. ii, p. 17.

Ericeæ, *Juss.*
23. Andromeda *tetragona*, L.—Hook. Fl. Bor. Am. ii. p. 38.

Monotropeæ, *Nutt.*
24. Pyrola *rotundifolia*, L.—Hook. Fl. Bor. Am. ii. p. 46.

Scrophularineæ, *Juss.*
25. Pedicularis *hirsuta*, L.—Hook. Fl. Bor. Am. ii. p. 109.

Amentaceæ, *Juss.*
26. Salix *Arctica*, Br.—Hook. Fl. Bor. Am. ii. p. 152.

MONOCOTYLEDONES.
Junceæ, *Juss.*
27. Luzula *hyperborea*, Br.—Hook. Fl. Bor. Am. ii. p. 188.

Cyperaceæ, *Juss.*
28. Carex *dioica*, L.—Hook. Fl. Bor. Am. ii. p. 208.
29. Carex *membranacea*, Hook. Fl. Bor. Am. ii. p. 220.
30. Carex *cæspitosa*, L.—Hook. Fl. Bor. Am. ii. p. 217.
31. Carex *ustulata*, Wahl.—Hook. Fl. Bor. Am. ii. p. 224.
32. Eriophorum *capitatum*, Host.—Hook. Fl. Bor. Am. ii. p. 231.

Gramineæ, *Juss.*
33. Hierochloe *alpina*, Ræm. and Sch.—Hook. Fl. Bor. Am. ii. p. 234.
34. Colpodium *latifolium*, Br.—Hook. Fl. Bor. Am. ii. p. 238.
35. Dupontia *Fischeri*, Br.—Hook. Fl. Bor. Am. ii. p. 242.
36. Poa *Arctica*, Br.—Hook. Fl. Bor. Am. ii. p. 246.
37. Poa *angustata*, Br.—Hook. Fl. Bor. Am. ii. p. 246.
38. Poa *alpina*, L.—Hook. Fl. Bor. Am. ii. p. 246.

SPECIMENS OF ROCKS,
Described by James Tennant, Esq.,
Professor of Mineralogy in King's College, London.

Cape Lady Pelly, 67° 30' N. 88° W.
    Gneiss.

Near Point Hargrave, 67° 25' N. 87° 35' W.
    Gneiss.

Cape T. Simpson, 67° 22' N. 87° W.
    Gneiss with chlorite.
    Mica-slate.
    Mica-slate, with indistinct crystals of precious Garnets.

Isthmus connecting Ross's Peninsula with the Continent.
    Felspar.

Simpson's Peninsula, 68° ⅓' N. 88° 20' W.
    Compact argillaceous Limestone.

A Hill on the western shore of Halkett's Inlet, 69° 14' N. 90° 50' W.
    Cellular Quartz, coloured by oxide of Iron.
    Mica-slate full of Garnets.

Helen Island, one of the Harrison Group in Pelly Bay, 68° 54' N. 89° 52' W.
    Felspar—red colour.
    Gneiss; the Felspar, Mica, and Quartz distinctly stratified.
    Gneiss; the Felspar red and greatly predominating.

Beacon Hill, near Fort Hope, 66° 32' N. 86° 56' W.
    Granite.
    Ditto, with a small quantity of Mica; the Felspar red, and constituting four-fifths of the mass.
    Gneiss, with veins of red Felspar running diagonally to the stratification.
    Mica-slate.

North Pole River.
    Mica-slate.
    Ditto, with veins of Quartz.
    Gneiss.
    Ditto, the Felspar red and greatly predominating.
    Ditto, the Felspar very friable.
    Quartz rock with Felspar.
    Argillaceous Limestone, compact.

North Pole Lake, 66° 40' N. 87° 2' W.
Gneiss.
Mica-slate.

Repulse Bay, 66° 32' N. 86° 56' W.
Quartz, coloured by oxide of Iron, and containing minute particles of Gold.

Melville Peninsula, 68° 27' N. 85° 24' W.
Hornblende-slate.

Munro Inlet.
Granite, the Felspar greatly predominating.

Island near the north point of Rankin's Inlet.
Quartz, enclosing chlorite and Copper Pyrites.
Talcose-slate.
Carbonate and silicate of Copper, with Copper Pyrites on argillaceous slate.
Ditto, with a thin coating of green carbonate of Copper.
Mica-slate.
Chlorite-slate, friable.
Ditto, with very thin veins of Calcareous Spar running diagonally in stratification.

Island near the south point of Rankin's Inlet.
Quartz and Iron Pyrites, the latter crystallized in cubes, the faces of which are not above one-sixteenth of an inch.
Quartz, with Iron Pyrites, and superficially coloured by oxide of Iron.
Hornblende-slate.
Mica-slate.
Chlorite-slate.

Dip of the needle and force of magnetic attraction at various stations along the west shore of Hudson's Bay, and at Fort Hope, Repulse Bay.

Name of Station.	Latitude N.	Longitude W.	Date.	Times.	Dip Mean.	Time of 10 Vibrations. Needle No. 2 deflected, 20 deg. from dip.	Therm.	Variation of Compass.
	deg. mi. sec.	deg. mi. sec.		h. mi.	deg. mi. sec.		deg. mi.	deg. mi. sec.
York Factory	57 0 2	92 26 0	5 Nov. 1845	9 0 A.M.	83 47 0		+31 0	
,,	57 0 0	92 26 0	8 ,,	9 0 ,,	83 43 0		+25 0	
,,	,,	,,	12 ,,	2 30 P.M.	83 37 0		+25 0	
,,	,,	,,	15 ,,	9 0 A.M.	83 41 0		+33 0	
,,	57 0 0	92 26 0	19 ,,	9 0 ,,	83 42 5		+25 0	
,,	,,	,,	22 ,,	9 30 ,,	83 43 4		+ 3 0	
,,	,,	,,	26 ,,	9 30 ,,	83 48 7		− 4 0	
,,	,,	,,	29 ,,	9 30 ,,	83 42 5		−13 0	
,,	,,	,,	3 Dec. ,,	9 30 ,,	83 54 2		− 6 0	
,,	,,	,,	6 ,,	9 30 ,,	83 43 2		+ 8 0	
,,	,,	,,	10 ,,	9 30 ,,	83 43 5		−19 0	
,,	,,	,,	13 ,,	9 30 ,,	83 48 2		0 0	

Name of Station.	Latitude N.	Longitude W.	Date.	Times.	Dip Mean.	Time of 10 Vibrations. Needle No. 2 deflected, 20 deg. from dip.	Therm.	Variation of Compass.
	deg. mi. sec.	deg. mi. sec.		h. mi.	deg. mi. sec.		deg. mi.	deg. mi. sec.
York Factory	57 0 0	92 26 0	17 Dec. 1845	9 35 A.M.	83 40 9		−11 0	
,,	,,	,,	20 ,,	9 30 ,,	83 39 1		−16 0	
,,	,,	,,	24 ,,	10 10 ,,	83 45 5		−23 0	
,,	,,	,,	31 ,,	10 30 ,,	83 46 0		+ 7 0	
,,	,,	,,	3 Jan. 1846	10 30 ,,	83 46 1		+20 0	
,,	,,	,,	7 ,,	10 30 ,,	83 47 0		+ 5 0	
,,	,,	,,	10 ,,	10 30 ,,	83 45 5		+ 7 0	
,,	,,	,,	14 ,,	10 30 ,,	83 43 9		− 2 0	
,,	,,	,,	21 ,,	10 30 ,,	83 44 8		−10 0	
,,	,,	,,	24 ,,	10 30 ,,	83 41 7		+23 5	
,,	,,	,,	28 ,,	10 30 ,,	83 45 8		+15 0	
,,	,,	,,	31 ,,	10 0 A.M. 3 0 P.M.	83 45 8		−15 0 − 3 0	
,,	,,	,,	4 Feb. ,,	10 0 A.M. 3 0 P.M.	83 50 5		−12 5 −14 0	
,,	,,	,,	7 ,,	10 0 A.M.	83 45 5		−11 5	

Name of Station.	Latitude N.	Longitude W.	Date.	Times.	Dip Mean.	Time of 10 Vibrations.	Therm.	Variation of Compass.
	deg. mi. sec.	deg. mi. sec.		h. mi.	deg. mi. sec.	Needle No. 2 deflected, 20 deg. from dip.	deg. mi.	deg. mi. sec.
York Factory......	57 0 0	92 26 0	11 Feb. 1846	10 0 A.M.	83 44 8		− 5 0	
,,	,,	,,	,,	3 30 P.M.			−11 3	
,,	,,	,,	14 ,,	9 30 A.M.	83 41 6		−23 0	
,,	,,	,,	,,	3 20 P.M.	83 38 1		+ 8 0	
,,	,,	,,	18 ,,	9 30 A.M.	83 36 6		+ 6 0	
,,	,,	,,	,,	3 30 P.M.			+ 3 0	
,,	,,	,,	21 ,,	9 30 A.M.	83 41 0		−11 5	
,,	,,	,,	,,	3 30 P.M.			+ 6 0	
,,	,,	,,	25 ,,	9 30 A.M.	83 40 9		−23 0	
,,	,,	,,	,,	3 30 P.M.			−10 5	
,,	,,	,,	28 ,,	9 30 A.M.	83 39 7		−13 0	
,,	,,	,,	,,	3 30 P.M.			+ 4 0	
,,	,,	,,	4 Mar. ,,	9 30 A.M.	83 44 1		+ 6 5	
,,	,,	,,	,,	3 30 P.M.			+ 4 0	
,,	,,	,,	7 ,,	9 40 A.M.	83 42 5		+29 0	
,,	,,	,,	,,	3 30 P.M.			+37 0	
,,	,,	,,	11 ,,	9 30 A.M.	83 44 6		+26 0	
,,	,,	,,	,,	3 30 P.M.			+25 5	
,,	,,	,,	14 ,,	9 30 A.M.	83 40 9		+12 0	
,,	,,	,,	,,	3 30 P.M.			+22 5	
,,	,,	,,	18 ,,	9 30 A.M.	83 39 6		+15 0	
,,	,,	,,	,,	3 40 P.M.			+21 5	
,,	,,	,,	21 ,,	9 30 A.M.	83 37 7		+ 2 8	
,,	,,	,,	,,	3 30 P.M.			+ 5 0	
,,	,,	,,	25 ,,	9 40 A.M.	83 47 0		+30 0	
,,	,,	,,	,,	3 30 P.M.			+30 5	
,,	,,	,,	28 ,,	9 35 A.M.	83 43 8		+ 8 0	
,,	,,	,,	,,	3 30 P.M.			+ 8 0	

Name of Station.	Latitude N.	Longitude W.	Date.	Times.	Dip Mean.	Time of 10 Vibrations. Needle No. 2 deflected, 20 deg. from dip.	Therm.	Variation of Compass.
	deg. mi. sec.	deg. mi. sec.		h. mi.	deg. mi. sec.		deg. mi.	deg. mi. sec.
York Factory......	57 0 0	92 26 0	1 April 1846	9 30 A.M.	83 42 8		+8 0	
"	"	"	"	3 30 P.M.			+15 0	
"	"	"	4 "	9 30 A.M.	83 45 2		+35 0	
"	"	"	"	3 30 P.M.			+25 0	
"	"	"	11 "	9 40 A.M.	83 40 6		+41 0	
"	"	"	"	3 30 P.M.			+42 5	
"	"	"	15 "	9 35 A.M.	83 35 7		—3 0	
"	"	"	"	3 30 P.M.			—6 0	
"	"	"	18 "	9 30 A.M.	83 40 2		+9 0	
"	"	"	"	3 30 P.M.			+29 0	
"	"	"	22 "	10 30 A.M.	83 38 9		+45 0	
"	"	"	"	3 35 P M			+40 0	
"	"	"	25 "	10 0 A.M.	83 35 5	Ther. +41° 0' 21s-34	+43 0	
"	"	"	"	3 30 P.M.			+32 0	
"	"	"	29 "	9 45 A.M.	83 38 0	Ther. +46° 0' 21s-23	+42 0	
"	"	"	"	3 30 P.M.			+43 0	
"	"	"	2 May "	9 30 A.M.	83 38 5		+39 0	
"	"	"	"	3 30 P M			+47 0	
"	"	"	6 "	9 30 A.M.	83 37 9	Ther. +66° 0' 21s-31	+51 0	
"	"	"	"	3 30 P.M.			+67 0	
"	"	"	16 "	9 35 A.M.	23 39 0	Ther. +43° 0' 21s-13	+36 0	
"	"	"	"	3 35 P.M.			+44 0	
Creek.............	58 2 0	92 20 0	20 June "	3 45 P.M.	84 46 4		+49 0	
Churchill	58 43 50	94 14 0	29 "	9 47 A.M.	84 50 8	Ther. +61° 0' 21s-14	+60 0	
"	"	"	"	3 35 P.M.			+61 0	
"	"	"	1 July "	10 30 A.M.	84 43 9		+88 0	
"	"	"	"	3 0 P.M.			+60 0	

Name of Station.	Latitude N.	Longitude W.	Date.	Times.	Dip Mean.	Time of 10 Vibrations. Needle No. 2 deflected, 20 deg. from dip.	Therm.	Variation of Compass.
	deg. mi. sec.	deg. mi. sec.		h. mi.	deg. mi. sec.		deg. mi.	deg. mi. sec.
Churchill	58 43 50	94 14 0	4 July 1846	8 10 P.M.	84 44 5		+41 0	
Knapp's Bay......	61 9 42	,,	8 ,,	10 45 A.M.	86 18 3	}	+52 0	
,,	,,	,,	8 ,,	3 0 P.M.			+51 0	
,,	,,	,,	12 ,,	5 15 P.M.	87 16 3	Ther.+54° 0' 20s-84	+58 0	
,,	,,	88 0 0	18 ,,	Noon.	86 36 5		+52 0	
Near Wager River	64 6 0	,,	21 ,,	4 5 P.M.	87 10 6	Ther.+65° 5' 21s-03	+54 0	
,,	65 10 0	87 10 0	22 ,,	11 35 A.M.		}	+52 0	
Repulse Bay......	65 15 36	,,	27 ,,	11 15 A.M.	88 16 7	Ther.+57° 5' 21s.7	+55 0	
Flett's Portage	66 32 0	,,	28 ,,	2 40 P.M			+57 0	
Descent Portage ..	,,	,,	31 ,,	3 15 P.M			+90 0	
Cape Lady Pelly ..	,,	,,	3 Aug. ,,	6 20 P.M.			+82 0	
3 Miles N.W. of do.	,,	,,	,,	6 50 P.M.			+53 0	
Fort Hope	,,	,,		5 30 P.M.	88 27 1	Ther.+52° 0' 21s.-8	+52 0	
	66 32 0	86 56 0	18 Nov. ,,	11 15 A.M.	87 51 5	}	— 6 0	West
,,	,,	,,	21 ,,	2 0 P.M.			— 5 0	62 50 30
,,	,,	,,		9 45 A.M.			+ 6 0	
,,	,,	,,		2 15 P.M.	88 11 4	Ther.+10° 5' 22s.-66	+10 0	

Name of Station.	Latitude N.	Longitude W.	Date.	Times.	Dip Mean.	Time of 10 Vibrations.	Therm.	Variation of Compass.
	deg. mi. sec.	deg. mi. sec.		h. mi.	deg. mi. sec.	Needle No. 2 deflected, 20 deg. from dip.	deg.	deg. mi. sec.
Fort Hope........	66 32 0	86 56 0	25 Nov. 1846	2 10 P.M.	88 8 9		−21	
”	”	”	5 Dec. ”	10 0 A.M.	88 13 9	Ther. + 9° 0′ 22s.-6	−15	
”	”	”	” ”	2 0 P.M.			−13	
”	”	”	12 ”	10 0 A.M.	88 13 3		−16	
”	”	”	” ”	2 5 P.M.			+6	
”	”	”	16 ”	10 0 A.M.	88 12 7		+8	
”	”	”	” ”	2 20 P.M.			0	
”	”	”	23 ”	10 0 A.M.	88 16 3		+2	
”	”	”	” ”	2 0 P.M.			+7	
”	”	”	2 Jan. 1847	10 10 A.M.	88 17 5		−8	
”	”	”	” ”	2 30 P.M.			−23	
”	”	”	10 Feb. ”	9 50 A.M.	88 10 9		−21	
”	”	”	” ”	2 10 P.M.			−22	
”	”	”	13 ”	9 50 A.M.	88 13 5		−20	
”	”	”	” ”	2 10 P.M.			−28	
”	”	”	17 ,”	9 50 A.M.			−26	
”	”	”	” ”	2 15 P.M.			−36	
”	”	”	24 ”	9 55 A.M.			−33	
”	”	”	” ”	2 10 P.M.			−22	
York Factory......	57 0 0	92 26 0	18 Sept. ”	9 15 A.M.	83 47 0		−22	
”	”	”	” ”	3 10 P.M.			+52	

Fort Hope, Repulse Bay.—*Abstract of*

Day of the Month.	Temperature of the Atmosphere taken eight times in twenty-four hours.			Prevailing Winds.	
	Highest.	Lowest.	Mean.	Direction.	Force.
	deg.m.	deg.m.			
1	+35	+27	+29.7	E.S.E	2—4
2	+37	+27	+31	E.S.E.	5—4
3	+36	+25	+31	E.—Vble.	9—1
4	+34	+28	+30.3	E. by S.	8
5	+42	+26	+32.7	O.—N.N.W.	0—7
6				N.	6
7	+31	+25	+27	N.	6
8	+35	+26	+30.5	N.N.W.	6
9				N.N.W.	6
10	+32	+30	+31.3	N.N.W.—O.—S.E.	4—5
11	+34	+31	+32.5	E. by S.	10—8
12				E. by S.—S. E. by E.	9—5
13				S.W. by S.—S.W.	5—9
14					
15	+45	+45	+45	S.S.E.	4
16	+34	+25	+28.7	Vble.—O.—E. by N.	1—2
17	+32	+24	+28	W.	2—3
18	+29	+26	+27.7	N.W.—W.N.W.	6—7
19	+33	+26	+29.7	W.N.W.—O.—E.	9—0
20	+32	+24	+28	N.N.W.	5—4
21	+36	+24	+29.3	N.—O.—E.	0 3
22	+31	+23	+27.7	N. by W.	5—6
23	+28	+16	+22.3	W.N.W.	3—4
24	+42	+21	+29.3	Vble.	1—0
25	+30	+16	+24.3	Vble.	0—2
26	+30	+26	+29	E.N.E.	8—9
27	+26	+24	+25	N. by W.	5—6
28	+26	+20	+22.7	N.N.W.	7—6
29	+24	+22	+23	W.N.W.	4
30	+22	+18	+19.7	Vble.—S. E. by E.	1—4
			714.4		
			+28.57		

Meteorological Journal for September, 1846.

Barometer and Thermometer attached.		Remarks on the Weather, &c.
Barom.	Thermo.	
....	....	c. c. o. Solar halo with parhelia.
....	....	c. c. c.
....	....	s. b. c.
....	....	c. c. c. p. of sleet.
....	....	c. c. o. Full moon.
....	....	p. s. o.
....	....	p. s. c.
....	....	c. p. s.
....	....	c. p. s.
....	....	c. b. c. o.
....	....	s. c. s. c. b. much drift.
....	....	o. c. c. ☾ last quarter.
....	....	b. c.
....	....	c. p. s.
....	....	c. c. c.
....	....	b. c.
....	....	o. s. s.
....	....	s. s.
....	....	s. o. c s.
....	....	c. c. c.
....	....	s. s. b. Aurora visible to the southward at 8 P M.
....	....	b. b. c.
....	....	o. b. c. o.
....	....	c. o.
....	....	s. s. s.
....	....	s. drifting.
....	....	p. so. drifting.
....	....	b. c.
....	....	h. b. s.

Fort Hope, Repulse Bay.—*Abstract of*

Day of the Month.	Temperature of the Atmosphere taken eight times in twenty-four hours.			Prevailing Winds.	
	Highest.	Lowest.	Mean.	Direction.	Force.
	deg.m.	deg.m.	deg.m.		
1	+27	+25	+26	Vble S.W.—N.W.	1—5
2	+25	+16	+21	N.W.	8
3	+24	+10	+18	Vble. E. by S.	1—5
4	+38	+38	+38	S.E. by E.	4
5	+37	+30	+33	E.	2—4
6	+33	+28	+30.3	N.E.	3—4
7	+30	+28	+29	N.E.	4—3
8	+28	+25	+26.3	N.—N.N.W.	4—5
9	+22	+21	+21.5	N.W.—O.—Vble.	3—0—2
10	+27	+26	+26.5	E.	8—9
11	+32	+28	+30	N.E.—O.	1—0
12	+27	+25	+25	N. by W.	7—9
13	+29	+27	+28.1	N. by W.	8—9
14	+26	+13	+23.2	N.	10—11
15	+12	+10	+11	N. by W.	10—11
16	+ 5	0	+ 2.6	N.N.W.	7—4
17	+ 3.5	— 1	+ 0.8	N.N.W.	7—8
18	+ 6	— 0.8	+ 1.7	S.W.W.—W.N.W.	4—6
19	+ 2	— 4.8	— 0.7	N.—N.N.W.	5—9
20	+ 3	— 2.5	— 0.3	N.W.	10—11
21	— 2.8	—10	— 6	N W.—N.W.byN.	7—11
22	— 4.5	—15	— 8.1	N.W. Vble. S.W.	0—2
23	+ 5.3	— 0.5	+ 3	N.W. by W.—N.W. by N.	3—5
24	— 0.	— 6.4	— 4.2	N.W.byW —N.W.	4—5
25	+ 4.5	— 6.2	— 1.8	N.W. by N.	5
26	— 7.3	—10.2	— 8.5	N.W.—N.W. by N.	4—6
27	— 6.	—15	—10.6	N.W. by N.—N.W.	0—3—5
28	— 1.8	—11.8	— 6.4	N.W. & N.N.W.	0—4
29	+10	+ 3.1	+ 8.4	S.S.E. S.—calm.	0—2—4
30	+25.3	+21	+23.4	S.S.E.—S.W.—W. by N.	2—8
31	+10	0	+ 5.2	S. N.W. W.S.W. N.N.W.	1—4
			389.4		
			+12.56		

Meteorological Journal for October, 1846.

| Barometer and Thermometer attached. || Remarks on the Weather, &c. |
Barom.	Thermo.	
....	....	s. ps.
....	....	b. c. drifting.
....	....	h. p. s. o. s.
....	....	h. p. r.
....	....	h. wet.
....	....	h. p. s. o. p. s.
....	....	h. p. s.
....	....	c. o. o.
....	....	h. c. c.
....	....	s. drifting.
....	....	s. s. s.
....	....	s. with much drift.
29.338	+49	s. and much drift.
29.431	+46.3	s. and drift.
29.690	+44	s. much drift.
29.605	+30.5	h. c. ; drift; haze and some drift—parhelia; haze with scaly snow; faint aurora to the S. and S. by E. alt. 12°.
29.719	+32.8	b. c., much drift; aurora to the S.S.E. parallel to the horizon; alt. 12°.
29.641	+31.5	b. c., drift; cirrus; some faint streaks of aurora to the W.
29.662	+29	b. c., drifting; solar halo with prismatic colours and parhelia; snow and much drift.
29.842	+29.5	s. much drift.
29.959	+30.5	b. c., much drift; at 8 P.M. several streaks of faint aurora extending across the zenith in a N.W. and S.E. direction; many rays in different parts of the heavens.
29.828	+28.5	
29.919	+32	f. o. f. o. s. o. s. b. c. f. s.
29.974	+31	b. c. o. drifting.
30.023	+29	o. drifting.
30.062	+29.3	o. m. b. c. drifting.
30.47	+26.5	b. c. m., some faint streaks of aurora in various parts of the sky bearing for the most part N.N.W. and S.S.E.
30.505	+26.	b. c., a few clouds near horizon; a very faint light yellow cloud aurora to the S.E. and N.W.
30.119	+30.3	c. s. b. c. s. o. m. b. c., cirrus extending from S.S.E. to N.N.W., resembling much the aurora. Lunar halo.
29.078	+39.7	o. m. o. s. b. c. o. drifting.
30.094	+34.3	b. b. c. c., solar halo; cirrus; 120 lunar distances were observed from Jupiter and at. Aquilæ, E. and W. of the moon. Lunar halo diam. 40° or 50°.

Fort Hope, Repulse Bay.—Abstract of

Day of the Month.	Temperature of the Atmosphere taken eight times in twenty-four hours.			Prevailing Winds.	
	Highest.	Lowest.	Mean.	Direction.	Force.
	deg.m.	deg.m.	deg.m.		
1	+18	− 3.0	+ 8.5	W.N.W. N.E. E.	2—7
2	+26.5	+22.3	+24.4	S.E. S.E. by E. E. by W.	2—5
3	+27	+25.5	+26.3	S.E. E.S.E.	2—5
4	+26	+21.5	+23.8	S.E.S. S.S.E.	3—5
5	+22	+ 0	+13.2	N. by W. N.W. by W.	2—7
6	− .5	− 9.5	− 3.5	W.N.W.	3—7
7	+11.5	+ 6	+ 9.7	N. by E.	4—7
8	+11	+ 5	+ 8.5	N.	4—7
9	+12.5	+ 9 5	+10.9	E.N.E. N.E.	3—10
10	+28.2	+22.5	+25.6	E.S.E. S. S.S.W.	3—8
11	+17	+ 2.5	+ 7.5	N.W. N.N.W. W. by N.	5—8
12	+ 2.3	− 8.5	− 1	N.N.E. W. N.N.W.	2—5
13	− 6	− 8	− 6.8	N. by W. N.N.W.	4—8
14	− 4.6	− 8.7	− 6.6	N.N.W. N. N. by W.	3—7
15	+ 4.5	−10.5	− 3.8	Calm. Vble. E.	0—4
16	+17.3	+15	+16.3	E. N.E. N.	1—6
17	+ 7.5	− 8	+ .25	N. by W.	4—6
18	− 4	− 9.2	− 7.1	N.W. by N. Calm S.W.	0—2
19	+21.7	+18	+20.61	S.S.E. S.E. E.	4—7
20	+12	− 8.8	+ 2.9	Calm. S. by E. N.	0—2
21	+ 4.5	− 4.2	− 0.9	S. S.E. E.	4—1
22	− 3	− 4.2	− 3.6	S. by E. W. N.W.	2—6
23	−18.5	−22.5	−19.77	N. by W. N.N.W.	3—5
24	−20.5	−25.2	−22.54	N.N.W.	5—1
25	−14.5	−24.5	−20.06	N.byE. N.W. N.W.byW.	1—3
26	−17.5	−23.5	−20.7	N.	6—9
27	−11.8	−15.5	−13.6	N. by W.	9—10
28	− 5.4	− 8.5	− 6.6	N. by W.	7—9
29	−16.5	−25.3	−20.3	N.N.W. W.N.W.	6—3
30	−17.5	−24.4	−21.	W. W.N.W. N.W.	6—3
			+20.59		
			+ 0.68		

Meteorological Journal for November, 1846.

Barometer and Thermometer attached.		Remarks on the Weather, &c.
Barom.	Thermo.	
30.011	+35	b. c. o. s. and drift.
29.715	+38	o. m s. o m. o s.
29.623	+38.7	o. m s. o s.
29.624	+39.5	o. m. b c. o m.
29.796	+41	o. m s. b c. b. drifting. A faint ray of aurora to the S. E. extending vertically towards the zenith.
30.009	+38.8	b. c. drifting. Some faint beams of aurora extending from S.W. to N.W., alt. 60°; one ray to the S.E. pointing towards the zenith.
29.894	+37.3	o. c. o. drifting.
30.1	+39.5	o. drifting.
39.996	+35.2	o. s. drifting thick.
29.598	+40.2	o s. o. b. c. o. much drift.
29.728	+38	o. s. o. m. b. c. drifting.
30.163	+38.1	b. c. m. b. drifting.
30.214	+34.9	b. m. b c. m. much drift.
30.39	+36.2	b. m. much drift. Solar halo and parhelia with prismatic colours; hazy near horizon; a faint beam of aurora to the westward directed toward the zenith; drifting.
30.239	+37	o. m. o. s.
29.963	+38	o. s. b. c. m. drifting.
30.102	+37	o. s. b. c. m. drifting. Three beams of aurora pointing towards the zenith; two of them bearing N.N.W., and the other S.E.
30.006	+33.7	b. c. f. o. m. At 9 A.M. there was a very red sky to the N. westward; sound heard at a great distance.
29.573	+36.7	o. s. b. c. drifting.
29.420	+36.8	o. s. m. o. s. f. b. c. m. At 7 h. 30 m. a faint aurora extending from W. to S.E., alt. 20°; motion rapid; no prismatic colours.
29.409	+37	o. s. b. c. s. o. f. s. b. m. s.
29.615	+39	b. c. Some faint streaks of aurora, most of them to the S. eastward, and pointed towards the horizon.
29.918	+33.7	b. m. b. c. Some faint rays of aurora visible this morning at 5 h. 30 m. in different parts of the heavens; drifting.
30.408	+33.7	b. c. drifting.
30.573	+30.8	b. b. m. Two faint beams of aurora bearing W.N.W. and pointing towards the zenith; altitude of lower limb 30°.
30.606	+32	b. m. b. much drift.
29.555	+31	b. m. o. s. drifting. Door drifted up.
29.41	+26.6	o. m. b. c. s. o. s. drifting.
29.894	+27.5	b. c. drifting.
30.354	+26	b. c. m. drifting.

FORT HOPE, REPULSE BAY.—*Abstract of*

Day of the Month.	Temperature of the Atmosphere taken eight times in twenty-four hours.			Prevailing Winds.	
	Highest.	Lowest.	Mean.	Direction.	Force.
	deg.m.	deg.m.	deg.m.		
1	—24	—27	—25.875	Calm. N.E. N.	0—3
2	—26.7	—30	—28.1	N.E. Calm. N.	1—0
3	—24.8	—28.5	—26.4	N. by W.	1—4
4	—24.8	—28	—29.97	N.W. by W. S.S.W.	4—0
5	—17.3	—21	—19.7	Calm. S. by E. S.S.E.	0—2
6	— 6.5	—11	— 9.14	E. by S. N.E. N.	5—2
7	—16.5	—24	—19.7	N.	5—7
8	—19.5	—25.6	—22.61	N.	9—8
9	+14	—15	+ .03	N.N.W. N.N.E. N.E.	11—5
10	+17	+14.8	+15.74	N.E. by N. N.E. E.	4—6
11	+12.7	+ 9.8	+11.6	N. by E. N.N.W. N.W.	4—1
12	+ 4	— 6	+ .74	S. S.S.E. Calm.	0—3
13	—13	—17	—14.93	N. N. by W.	4—1
14	—19	—23	—20.94	Calm. Vble.	0—2
15	— 9	—19	—16.55	N.N.W. N. by W.	1—4
16	0	— 3	— 1.64	N. E.N.E. Calm. Vble.	0—1
17	— 5	— 9.6	— 6.05	Vble. W.N.W.	1—2
18	— 6	— 8.5	— 7.04	N. by W. W. Vble.	2—1
19	—14.2	—20	—17.4	N. by W. N.N.W.	5—4
20	— 8.7	—13	—10.56	S. by W. N. by W. N.	1—4
21	—20.7	—32.3	—24.83	N.W. Vble. N.	1—2
22	—30.5	—36.5	—33.4	W. Calm. N. by E.	0—2
23	—21.4	—26	—23.3	N.N.E. N.E.N.	0—1
24	—31	—35.3	—33.13	N.	7—10
25	—36	—38	—36.83	N. by W.	10—8
26	—34	—38	—36.46	N. by W. N.	8—11
27	—30	—30	—30	N.	10—11
28	—30.8	—34.8	—33.01	N. N. by W.	6—4
29	—24.5	—40	—35	N.W. by W. Vble. N.N.W.	0—5
30	—25	—32.3	—29.63	N.	6—9
31	—23	—32.5	—29.25	N. by W. Vble. N.	1—7
			597.43		
			—19.27		

Meteorological Journal for December, 1846.

Barometer and Thermometer attached.		Remarks on the Weather, &c.
Barom.	Thermo.	
30.452	+18.75	b. c.
30.237	+19.6	b. c. b. c. m. Lunar halo.
30.886	+16.3	b. c. b. c. m.
30.013	+17	b. c. m.
29.778	+17.6	b. c. m. parhelia with prismatic colours; aurora visible to the south in two arches arising from near the horizon to the zenith.
29.480	+27.5	o. s. b. c.
29.764	+26	b. m. c. drifting.
30.039	+23	b. c. drift.
29.974	+22	s. o. drifting.
29.892	+28.3	s. o. b. c. o. s. drifting.
29.759	+32	o. s. m.
30.016	+26.6	o. m. s. b. m.
30.36	+31	b. m. b. c. The sky to the north had a beautiful lake coloured tint at sunset; the most brilliant display of aurora I have observed this winter, the centre being towards the true south, and gradually rising from an altitude of 12° to 70° or 80°. It was of a pale yellowish green colour. Horizontal needle not affected.
30.473	+26	b. c. m. Some faint beams of aurora in different parts of the heavens. A very faint aurora to the southward.
30.37	+27	b. m. b. c. o. A very faint aurora; centre true south.
30.186	+30.7	o. m.
30.205	+27.6	o. m. b. m. Wind variable from N. to E.; faint aurora to the S.; alt. 10°; centre S.S.W. 30°.
30.274	+29.3	o. b. c. m. Aurora faint to the S. by W.
30.245	+27.3	b. c. m. drifting.
30.259	+28	b. c. o. s.
30.268	+29	b. m. Arch of aurora across zenith nearly east and west; brightest at western extremity.
30.264	+22.3	b. c. b. m.
30.168	+25.3	b. m. b. c. b. m. s. Speculæ of snow falling. Lunar halo faint.
30.065	+23.6	b. m. much drift.
29.996	+22	b. m. much drift.
29.83	+20	b. c. m. much drift.
29.523	+15.5	b. c. m. much drift.
29.536	+14.3	b. m. b. drifting.
29.603	+14.3	b. b. c. A faint halo, centre S., alt. about 20°; wind variable from N. to W. by S.; cirrus clouds; halo round moon.
29.577	+11.6	b. c. drifting; much drift.
29.564	+15.3	b. c.

FORT HOPE, REPULSE BAY.—*Abstract of*

Day of the Month.	Temperature of the Atmosphere taken eight times in twenty-four hours.			Prevailing Winds.	
	Highest.	Lowest.	Mean.	Direction.	Force.
	deg.m.	deg.m.	deg.m.		
1	—23.5	—32	—26.96	N.N.W. N.W.byW. N.byW.	1—6
2	—29.5	—33.5	—31.8	N.N.W. N. by W. N.W.	2—5
3	—30.3	—32	—31.4	N. by W. Calm. N.N.E.	0—1
4	—31	—34	—32.82	N. Calm. N.	0—2
5	—27.5	—30	—28.61	N. ½ W.	5—8
6	—26.5	—31	—28.3	N.N.W.	6—8
7	—40	—42	—40.9	N.W. Calm. W. N.W.N.	0—1
8	—44	—47	—46.7	N.W. N.N.W. N. by W.	1—7
9	—38	—40	—39	N.	10—11
10	—12	—17	—14.5	N.N.W.	10—12
11	—10	—10	—10	N. by W.	7—11
12	—12	—16	—14	N. by W.	7—8
13	—28.5	—33.5	—30.8	N.N.W. N. by W.	6—7
14	—33.8	—36.3	—35.1	N.by W. N.½ W. N. by W.	7—5
15	—38	—39.5	—38.7	N. by W. N.W. N.N.W.	2—5
16	—39.3	—41	—37.07	N. by W. N.N.W. N.by W.	2—6
17	—38	—41	—39.6	N. by W.	7—8
18	—37	—40	—38.95	N.W. by N. N.by W.	2—4
19	—25	—31	—30.6	N.N.W. N.N.W.	9—11
20	—14	—20	—17	N.N.W.	8—10
21	—20.5	—26.5	—23.4	N. by W. N.N.E. N.	2—9
22	—14	—26	—18.87	N.W. N.N.W.	6—11
23	—10	—13	—11.2	N.N.W.	9—11
24	—13	—13	—13	N.N.W.	9—11
25	—26.5	—32.5	—29.25	N.N.W.	4—7
26	—31.5	—37	—34.47	N. Calm. Vble. N.	0—1
27	—29	—35	—32.05	N. N. by W.	1—2
28	—33.3	—35.5	—34.65	N. by W.	6—7
29	—36	—42.7	—39.25	N. by W. W.N.W. N.W.	4—1
30	—24.7	—36.5	—28.64	S. by W. Vble. E.	1—5
31	—27.5	—35	—31.5	N. by W.	4—7
			909		
			—29.32		

Meteorological Journal for January, 1847.

Barom.	Thermo.	Remarks on the Weather, &c.
29.908	+17	b. c. b. c. s. drifting.
30.128	+16	b. m. b. Faint aurora, centre S.W. by S., alt. 15°; drifting; some streaks of aurora to the southward pointing to the zenith.
30.134	+18.5	b. c. b. Much refraction; thermometer in house +11°; a beam of aurora to the south pointing to the zenith.
30.023	+15.6	b. b. Hills much refracted; aurora faint; centre of arch S. by W.; alt. 10°; aurora in a narrow line parallel to horizon, alt. 4°, extent 70°, centre south.
29.93	+14.6	b. c. m. drifting.
30.04	+14.6	b. m. drifting. A faint aurora extending from S.S.E. across the zenith.
29.861	+12.6	b. c. m. Mercury froze after two hours' exposure.
29.8	+11	b. b. drifting.
29.974	....	Much drift; could not get out to see thermometer, door being drifted up.
29.139	+ 6	o. o. Much drift; obliged to take the thermometers into the house, as the pillars of snow on which the posts were placed were nearly all blown away.
29.193	+10.5	o. b. m. Much drift; a beam of aurora S.E.; alt. 25°.
29.309	+14.5	b. m. Much drift; very faint aurora; centre W. by N.; alt. 10°.
29.549	+12.3	b. m. drifting; a very faint aurora, centre S.S.W., alt. 16°; extent 60° or 70°.
29.588	+13	b. c. m. drift; arch of aurora faint, alt. 11°, centre S.S.W., extent 90°.
29.608	+ 7.6	b. m. c. Streams of bright light shooting from the sun to the alt. of 5°.
29.67	+ 7	b. c. b. drifting, stratus; arch of aurora faint, centre south, alt. 18°, extent 60°. Centre S.S.W., alt. 12°, extent 90°.
29.887	+13	b. m. drifting. Aurora visible, faint but brightest to the westward; centre S., alt. 60°.
29.245	+ 6	b. c. b. c. m. A very faint arch of aurora from N.W. by N. extending across zenith.
29.662	+ 7	m. o. much drift; door drifted up.
29.472	+11	o. q. much drift.
29.604	+ 9.5	b. m. much drift.
29.445	+ 8	b. m. o s. o. m. q. s. o. q. drifting.
29.273	+ 9.5	o. m. much drift.
29.366	+10	o q. gale all night; much drift.
29.83	+ 8	b. m. drifting; solar halo with parhelia.
30.035	+ 6.3	b. A faint arch of aurora across zenith S.W. and N.E.
29.911	+ 4.6	b. c. b. c. s. o. m. o. s.
29.908	+ 7.3	b. m. drifting. Very cold to the sensation; spiculæ of snow falling; a broad band of aurora, the lower edge having a reddish or lake tint, running parallel to the horizon; alt. 2°, centre S.W., extent 70°; some beams of aurora S.E. pointing towards the zenith.
29.954	+ 7.3	b. m.
29.737	+ 5.6	o. b. c. m. s. b. c. s.
29.714	+ 8	b. c. m. Cirrus; drifting.

Fort Hope, Repulse Bay.—Abstract of

Day of the Month.	Temperature of the Atmosphere taken eight times in twenty-four hours.			Prevailing Winds.	
	Highest.	Lowest.	Mean.	Direction.	Force.
	deg.m.	deg.m.	deg.m.		
1	—29.8	—38.5	—33.65	N.N.W. N.W. W.	6—1
2	—30.8	—37.3	—33.73	N.W. Vble. W. Calm. N.	0—1
3	—29	—35	—31.53	S.W. Calm. Vble.	0—1
4	—19	—26.5	—22.67	Calm. Vble. Calm.	0—1
5	—14	—20	—16.71	N.W. by S.	4—6
6	—14.7	—22.5	—17.5	N.	3—6
7	—22.5	—27	—25.16	Calm. N. by W. Calm.	0—1
8	—22.3	—30.5	—26.25	N. by W. N.N.W.	1—4
9	—20	—25.5	—21.65	N.W. N.W. by W.	1—6
10	—20	—27	—23.35	N. Vble. N. by W.	0—2
11	— 8.7	—18.3	—11.64	W.N.W. N. by W.	1—6
12	—18	—23.5	—20.25	N. by W.	8—6
13	—35.3	—38	—36.83	N.N.W. N. by W.	7—2
14	—26	—36.5	—31	N.W.	6—3
15	—37.5	—42	—39.83	N.	4—7
16	—36.5	—42	—39.14	N. by W.	7—5
17	—35.5	—40.5	—38.4	N. N. by W. N.W.	7—3
18	—27.5	—34.5	—30.57	N. N. by W. N.N.W.	1—7
19	—22	—32.5	—27.57	N. Vble. S.S.E.	4—1
20	—22.5	—27.5	—25.3	N. by W. N. N.N.W.	7—4
21	—19.5	—27	—22.83	N.N.W. N. S.E.	3—1
22	—13	—26.5	—18.85	N.E. N.N.W.	1—5
23	—23.5	—31.5	—26.57	N.N.W. N.	3—1
24	—23	—34.5	—27.43	W. W. by N. N. N.W.	1—4
25	— 9.5	—27.5	—20.2	W. Calm. Vble.	1—0
26	— 9.3	—22	—13.5	S.E. E. E. by N. N.	1—2
27	—24	—27.5	—25.54	N.W. by N. N.N.W.	4—6
28	—34.5	— 40	—39.2	N.N.W. N.W. by W.	6—3
			746.85		
			—26.68		

Meteorological Journal for February, 1847.

Barometer and Thermometer attached.		Remarks on the Weather, &c.
Barom.	Thermo.	
29.901	+ 7.6	b. m. b. q. drifting.
30.023	+ 5.3	b. b.
30.593	+ 2.6	b. c. o. cirrus and cirro-stratus.
30.219	+ 5	b. c.
30.339	+ 5.6	b. c. q. much refraction; drifting.
30.18	+11.	b. c. m. b. c. drifting.
30.' 4	+12.	b. c. cirrus; cloudy near horizon.
30.418	+10.3	b. m. spiculæ. much refraction.
30.432	+12.	o. m. b. c. m. drifting; solar halo with parhelia; a faint arch of aurora.
30.065	+ 8.3	b. c. cirrus; some faint beams of aurora south and south-south-west (say south-west).
29.865	+12.6	b. c. m. o. s. b. c. s. drifting.
29.71	+12.	b. m. much drift.
29.644	+10.5	b. m. b. drifting.
29.65	+10.	b. m. b.
29.816	+12.6	b. b. m. b. drifting.
29.899	+13.3	b. m. b. much drift.
29.84	+ 7.6	b. m. b. drifting.
29.869	+ 7.3	b. c. o. b. c. m. much drift.
29.9	+ 6.7	b. c. s. o. m. Solar halo with prismatic colours and parhelia.
29.9	+ 8	b. m. b. drifting.
30.329	+ 7	b. c. b. c. m.
30.276	+ 9.6	b. m. b. c. s. o. s. b. c. s. drifting.
30.459	+ 9.3	b. m. b. c. cirrus; Venus visible for the first time, the horizon having been too hazy to see her sooner.
30.326	+ 7	b.
30.008	+ 6	b. b. c. much refraction.
30.221	+ 8.3	b. m. c. b. c. s.
30.146	+12	b. m. c. b. c. s. b. c. m. drifting along the ground.
30.073	+11	b. m. drifting.

FORT HOPE, REPULSE BAY.—*Abstract of*

Day of the Month.	Temperature of the Atmosphere taken eight times in twenty-four hours.			Prevailing Winds.	
	Highest.	Lowest.	Mean.	Direction.	Force.
	deg.m.	deg.m.	deg.m.		
1	—30.5	—45	—37.5	N.byW. Chble. N.W.byN.	0—2
2	—30.5	—40.5	—35.4	N.W. by N. N.N.W.	2—4
3	—30	—37	—33.7	N.W. by N. N.N.W.	4—7
4	—27	—38	—32	N. by W. N.W. by N.	4—7
5	—26	—33	—28.4	N. by W. N.W. by N.	8—6
6	—27	—33	—29.4	N. by W.	8—4
7	—27.5	—37	—33	N.N. $\frac{1}{2}$ E.	7—5
8	—25	—31.5	—27.5	N. N. by W. N.N.W.	7—9
9	—20	—30.5	—25.3	N.N.W. N.W. by N.	4—2
10	—21	—33.5	—27.2	N.W. N.N.W.	1—4
11	—10.7	—27.5	—20	N.W. by N. N. by W.	1—3
12	—19.5	—30.5	—23.7	N.N.W. N. N. by W.	8—10
13	—15	—19.5	—16.5	N.N.W.	10—12
14	—13.5	—15	—14.5	N. by W.	11—7
15	—11	—19	—14.2	N. N.N.W.	8—5
16	—7.7	—19	—11.7	N.W. by N. N. by W.	3—6
17	—24	—30	—26.5	N. W.N.W. W.	1—6
18	—18.7	—37.5	—29.1	Calm. S.S.E. W.	0—6
19	—14	—29.5	—21.4	W. Vble.	2—1
20	—23.5	—32.5	—29.1	N.N.W. N. N. by W.	6—4
21	—23	—29.5	—25.9	W.N.W.	10—7
22	—16	—27	—21.6	N W. by N. W.	6—1
23	—16	—33	—22.6	N.W. Chble. N. by W.	1—6
24	—29	—33.5	—30.9	N. by W. N.N.W.	9—7
25	—27	—35	—30.4	N. by W. N.N.W.	7—9
26	—26.5	—35.5	—30.6	N. by W.	6—8
27	—24.5	—34	—28.1	N. by W. N.N.W.	6—8
28	—26	—35	—30.2	N. by W.	2—7
29	—22	—33	—26.37	N.N.W. N. W.N.W.	8—5
30	—15	—32	—20.54	N.W. N. by W.	2—6
31	—6	—14	—8.6	N.N.W. N.W. by N.	7—6
			811.91		
			—28.1		

Meteorological Journal for March, 1847.

Barom.	Thermo.	Remarks on the Weather, &c.
30.152	+ 4.3	b. b.
30.296	+ 4	b.
30.268	+ 4.6	b. m. drifting. The wind between noon and 2 P.M. went round for a few minutes, and then went back to its old direction.
30.399	+ 6.3	b. m. drifting.
30.492	+ 7	b. m. b. c. m. much drift.
30.63	+11.3	b. c. m. drifting.
30.514	+10.5	b. m. drifting.
30.232	+ 7.6	b. c. m. much drift.
30.194	+ 8	b. b. c.
30.179	+ 4	b. b. c. cirrus.
30.305	+ 4.7	b.
30.449	+ 9.7	b. m. much drift.
30.089	+ 7	b. q. thick drift.
30.07	+ 5	b. m. q. b. c. m. much drift.
30.886	+13	b. c. m. q. b. c. m. o. m. drifting.
29.578	+12	o. s. b. c. s. b. c. drifting.
29.814	+ 6.6	b. c. b. q. drifting.
29.99	+ 4.6	b. c. m. Solar halo with prismatic colours; drifting.
30.001	+ 5.6	b. m. b. c. cirrus.
29.569	+ 8	b. m. b. c. m.
29.372	+ 3	o. s. o. m. b. m. drifting.
29.673	+ 5	b. c. m. q. cirrus.
29.823	+ 6.7	b. c. m. o. s. Spiculæ; halo with prismatic colours; drifting.
29.854	+ 3.7	b. m. b. c. m. much drift; door drifted up.
29.899	+ .7	b. m. c. m. much drift; door drifted up.
30.196	+ 1.3	b. c. m. drifting.
30.046	− .3	b. m. b. c. m. drifting.
30.161	+ 1	b. m. c. drifting.
30.142	+ 2	b. m. drifting.
30.182	+ 3.5	b. c. m. o. m. drifting.
30.867	+10.6	b. c. m. b. c. s. o. s. drifting.

FORT HOPE, REPULSE BAY.—*Abstract of*

Day of the Month.	Temperature of the Atmosphere taken eight times in twenty-four hours.			Prevailing Winds.	
	Highest.	Lowest.	Mean.	Direction.	Force.
	deg.m.	deg.m.	deg.m.		
1	— 6.5	—18.3	—11.57	N.W. by W. W. by N.	3—6
2	— 0.5	—21	— 9.03	W. N.N.W. N.W.	2—4
3	+ 8	—23.5	— 6.7	Vble. Calm.	1—0
4	0	—13	— 4.5	N.W. by N. N.	2—1
5		—10		N. by W.	5
6	+11	—20	— 5.3	S.	4
7	+18	— 9	+ 3.67		
8	+20	— 2	+ 8.3		
9	+ 2	—12	— 5	N.N.W.	
10	+19	—15	+ 3.66	E.	
11	+10	—15	— 1.6	E.	
12	+16	—17	— 2	S.	
13	+21	—11	+ 5.3	N.N.W.	
14	+15	0	+ 6.6	W.	
15	— 7	—17	—11.3	N.N.W.	9
16	—10	—19	—15.3	N.	9
17	— 8	—22	—16.3	N.	
18	— 2	—20	—12	N.W.	
19	— 5	—25	—13.7	N.N.W.	....
20	— 5	—20	—12.67	N.	....
21	0	—22	—10.3	N.N.W.	
22	— 8	—22	—13.3	N. by W.	....
23	+17	—12	+ 1.67	Vble.	2
24	— 6	—10	— 4.3	N.W.	....
25	+ 7	— 2	+ 1	N.	
26	+ 5	—10	— 1.6	N.N.W.	....
27	+ 8	— 5	+ 2	N.N.W.	....
28	+10	— 3	+ 4	N.N.W.	....
29	+11	— 1	+ 4	N.N.W.	....
30	+20	— 1	+ 9.6	N.	....
			122.57		
			— 3.95		

Meteorological Journal for April, 1847.

Barom.	Thermo.	Remarks on the Weather, &c.
29.83	+10	b. c. m. drifting.
29.709		b. b. c.
29.708	+ 4	b. b. c. Barometer not registered after this. Thermometer with colourless rose to 5° only, although freely exposed to the sun's rays. At 6 P.M. a faint aurora of an orange colour; centre south; alt. 5°.
....	....	o. m. b. c. s. os.
....	....	o. s.
....	....	much drift all day.
....	....	much drift.
....	....	much drift and snow.
....	....	thick drift and snow. Some partridges seen.
....	....	drifting.
....	....	drifting thick.
....	....	snow and drift.
....	....	drifting.
....	....	drifting.
....	....	drifting.
....	....	drifting.

Fort Hope, Repulse Bay.—*Abstract of*

Day of the Month.	Highest.	Lowest.	Mean.	Direction.	Force.
	deg.m.	deg.m.	deg.m.		
1	+20	+ 4	+11.6	W.	....
2	+20	+ 5	+12	N.	....
3	+17	+ 4	+ 9.3	N. by W.	....
4	+10	+ 0	+ 3.3	N.N.W.	....
5	+10	− 4	+ 3.67	N.N.W.	....
6	+23	0	+ 9.3	Vble. Calm.	1—2
7	+24	− 1.5	+10.5	S.E. E.	2
8	+23	+ 6	+14.8	Vble. E. S.S.E.	1—3
9	+26	+16	+18.5	S.E. E.	2—6
10	+19.5	+12	+15.67	E. by S. E.N.E.	6—10
11	+32.3	+18.5	+24.6	S. by E. S.W. W.N.W.	1—6
12	+25.5	+10	+15.93	N.W.	2—6
13	+25	+ 4.5	+11.5	W.	7—6
14	+33	+18	+23.3	S.W.	
15	+17	+10	+12.67	N.	....
16	+15	+ 9	+11.3	N.W.	....
17	+20	+15	+17	W.N.W.	
18	+30	+15	+21.67	N.W.	
19	+40	+18	+27.6	S.	
20	+37	+21	+27.3	N.	
21	+28	+18	+21.3	N.	11
22	+22	+16	+18.3	N.	10
23	+25	+16	+21	N.	10
24	+33	+26	+28.66	N.E.	
25	+43	+23	+30.67	N.E. by N.	
26	+31	+24	+27.67	N.N.E.	
27	+28	+21	+24.66	N.	....
28	+25	+16	+20	N.W.	
29	+45	+18	+28	S.	
30	+43	+24	+30.67	S.E.	....
31	+23	+18	+21	N.	....
			553.44		
			+17.88		

Meteorological Journal for May, 1847.

Barometer and Thermometer attached.		Remarks on the Weather, &c.
Barom.	Thermo.	
....	....	Newman's improved Cistern Barometer used.
....	....	{Correction for capacities — $\frac{1}{51}$ Neutral point —30.302. Capillary action +.042. Temperature +60°.
....	....	A snow bird seen.
....	....	drifting.
....	....	drifting.
....	....	b. c.
....	....	o. s b c. s.
....	....	o. s. An inch of snow fallen.
....	....	o. s. o. o.
....	....	o. s and drifting thick.
....	....	o. s. pools of water. Beautiful evening.
....	....	b. c. drifting.
....	....	b. c. o. m.
....	....	fine weather.
....	....	thick weather.
....	....	Much snow drift.
....	....	Much snow and snow drift.
....	....	Much snow drift.
....	....	Snow and drift until evening.
....	....	Cloudy with snow.
....	....	Strong gale with drift.

Fort Hope, Repulse Bay.—Abstract of

Day of the Month.	Temperature of the Atmosphere taken three times in twenty-four hours.			Prevailing Winds.	
	Highest.	Lowest.	Mean.	Direction.	Force.
	deg.m.	deg.m.	deg.m.		
1	+25	+12	+19.3	S.	
2	+35	+17	+25.3	N.	....
3	+26	+14	+20	N.	
4	+32	+14	+21.7	N.W.	
5	+29	+18	+22	N.W.	
6	+43	+21	+28.3	Vble.	
7	+28	+18	+22	N.	
8	+30	+16	+22.7	N.	
9	+38	+24	+30.6	N.N.W. and Vble.	3—5
10	+39	+26	+31.3	N. and N.N.E.	1—3
11	+34	+28.5	+30.8	Vble. N.	1—6
12	+35	+26.5	+30.7	N. by W.	6—8
13	+37	+27	+32.3	N.	5—7
14	+40	+29.5	+34	N. by E.	2—4
15	+43.5	+26	+35.5	E. Vble. S.W.	2—3
16	+39.5	+36	+37.3	N. N.W.	4—2
17	+37	+30.5	+34	E. by S. S.E.	3—1
18	+38.5	+32.5	+34.67	E. N.E.	2—5
19	+34.5	+31	+32.5	N.N. by W.	7—9
20	+37	+33.5	+34.8	W.N.W.	10—11
21	+45.5	+33	+37.66	W. by N. S.E.	9—6—5
22	+40.5	+32	+35.1	N. N.N.W. N.W.	8—7
23	+42	+32.5	+36.2	W·N.W.	6—4—2
24	+46.5	+33	+38.73	Calm. Vble. S E.	0—2
25	+36.7	+32.5	+34.23	E. by S.	3—4
26	+37	+31.3	+33.66	E.S E. E. by N. N.E.	6—9
27	+34.3	+31	+32.6	N.W. W.N.W.	10—11
28	+34	+31.5	+32.83	W. W. by N. W.N.W.	9—8
29	+37.3	+33.7	+35	N.W. N.W. by W.	10—8—0
30	+41	+32.3	+35.6	W.N.W. N.W. N.	7—8
			942.51		
			+31.38		

Meteorological Journal for June, 1847.

Barometer and Thermometer attached.		Remarks on the Weather, &c.
Barom.	Thermo.	
....	....	A strong gale.
....	....	b. c. m. Arrived at the house from our journey at 8h. 20m. A.M. by watch, or 7h. 20m. true time.
....	....	b. c.
....	....	o. s.
....	....	o. s.
....	....	o. p. s.
....	....	o. b. c.
....	....	b. c. p. sleet.
....	....	b. c.
....	....	b. c. p. o. r. First rain this spring.
....	....	o. r. o. f. o. r.
....	....	s. o. r. o.
29.480	+37	p. r. b. c. b. c. p. r. b. c.
29.817	+49	b. c. q. o. r.
30.289	+40	o. b. c. p. s. Showers of snow and sleet during the night.
30.14	+40.3	o. b. c. Saw sun at midnight, lower limb touching the high ground.
30.147	+46.5	b. c.
30.04	+40	o. o. f. A few flakes of snow falling.
29.68	+38.7	o. s. o. w. s. Half inch of snow during the night. Wet snow.
29.273	+37	o. s. o. p. s. q. From 6 to 8 inches of snow during the night.
29.39	+35.6	b. c. q. o. s. q.
29.488	+40	o. p. s. q. b. c. q. b. c. p. s.
29.61	+38	o. s. b. c. p. s. q. b. c. p. r. q. Wet snow.

Fort Hope, Repulse Bay.—*Abstract of*

Day of the Month.	Temperature of the Atmosphere taken three times in twenty-four hours.			Prevailing Winds.	
	Highest.	Lowest.	Mean.	Direction.	Force.
	deg.m.	deg.m.	deg.m.		
1	+39	+29.3	+33.6	N.N.W. N. by W. N.	4—6
2	+38	+31.3	+34.6	N. N.W. by N. N.W.	7—4
3	+46.5	+32	+38.17	W. Calm.	7—6—0
4	+35.5	+33	+34.1	N.E.	6—5—4
5	+45.5	+35	+39.8	W.	5—3—6
6	+46	+34	+39.17	W.N.W. N. by W. Chble.	7—0
7	+49	+38	+43	E. by S. S.E. Calm.	2—4—0
8	+51	+35	+42	E. E.S.E. E.	3—5—1
9	+48.7	+32.3	+38.7	N. Vble. E.	5—2
10	+41	+35	+37.17	E.S.E.	5—6
11	+36	+33	+34.5	E. by N. Calm.	4—3—0
12	+39.3	+35	+36.7	N. N. by E.	3—5—6
13	+38	+33.5	+35.6	N. by W. N.	8—9
14	+38	+33.7	+35.23	N.	9
15	+42.5	+34	+37.2	N. by W.	9—10
16	+39	+35.3	+37.7	N. Calm.	10—7—0
17	+46	+36	+42.5	N.N.W. W. by N.	8—5—3
18	+43	+35	+39.5	Vble. Calm.	3—4—0
19	+47.3	+36	+41.6	N.W.	5—6—3
20	+55.5	+41	+46.9	N.N.W. N.W. Calm.	3—5—0
21	+57	+44	+49.17	N. Vble. N.N.W.	4—1—3
22	+47	+40	+42.5	Calm. N.N.W.	0—6—5
23	+49.3	+38.5	+43.26	N.N.W. N. N. by W.	8—7—8
24	+48	+36.5	+41.9	N. N.W. by N.	9—7—3
25	+52	+36	+43.16	N.W. Calm.	6—4—0
26	+43	+38	+40.2	S.S.E. E.S.E. E.	2—6
27	+51.5	+40	+44.17	N.E. Calm.	5—3—0
28	+60	+45	+51.8	W. W.N.W. W. by S.	2—3—2
29	+53.5	+47	+50.2	N.	4—3—1
30	+55	+38.3	+46.6	W. by N. N.	4—8—10
31	+48	+37.5	+42.5	N. by W.	3—8—5
			1285.4		
			+41.46		

Meteorological Journal for July, 1847.

Barometer and Thermometer attached.		Remarks on the Weather, &c.
Barom.	Therʼmo.	
29.786	+39.83	b. c. p. s. a little frost during the night.
29.838	+35.5	b. c.
29.986	+46	b. c. a beautiful night.
29.864	+40.3	o. p. o. f. p. r. o. sleet.
30.015	+43	b. c.
30.124	+42	b. c. b. c. q. Ther. at midnight +35°; coat of ice on pools where there is snow.
30.216	+49.5	b. c.
30.185	+46	b. c.
30.216	+40.3	o. b. c. o.
30.024	+42	o. b. c. o.
29.828	+42	p. r. f. o. f. w. o. Heavy rain during the night; wet fog and showers of rain.
29.802	+40	o. f. p. r. o. w. f.
29.938	+39	o. f. p. r. o. f. o. p. r. q.
29.968	+41.3	r. o. b. c. o.
29.905	+41.7	o. b. c. o. r. A great quantity of water coming down N. Pole River this morning; sleet.
29.865	+44.2	p. w. s. q. o. s. b. c. Snow showers all night; ther. at 6 p.m. +45°.
29.902	+47.2	o. b. c. at 5 p.m. Ther. at +54°.
		b. c. b. c. o.
29.716	+48	b. c. q.
29.714	+56	b. c.
29.776	+54.5	b. c.
29.794	+46.5	o. b. c. p. r. b. c.
29.791	+46	d. r. b. c. p. r. b. c.
29.858	+45.5	b. c.
29.967	+53	b. c.
29.815	+47.2	b. c. b. c. q.
29.917	+49	b. c.
30.038	+53.5	b. c.
30.113	+56.8	b. c.
30.017	+49	b. c. p. r. The barometer fell some hundredths lower than when registered at 6 a.m., but immediately began to rise as soon as the wind changed to the north.
30.102	+51.5	b. c.

FORT HOPE, REPULSE BAY.—*Abstract of*

Day of the Month.	Temperature of the Atmosphere taken three times in twenty-four hours.			Prevailing Winds.	
	Highest.	Lowest.	Mean.	Direction.	Force.
	deg.m.	deg.m.	deg.m.		
1	+52	+40	+44.8	N.	4—6—3
2	+56	+40	+47.7	N.N.W.	6—2—1
3	+49	+44.5	+46.2	N.W. N.N.W.	6—7—5
4	+41	+34.7	+36.9	N.N.W. N.	9—8
5	+54	+34	+62.5	N. N. by W.	7—6—3
6	+50	+46.5	+49.8	Vble. W.S.W.	3
7	+59.3	+43.5	+49.3	S.W. Calm.	4—5—0
8	+49.5	+42	+45.5	Vble. N.W.	1—2—6
9	+44.5	+37.	+39.83	N. N.W.	8—6—4
10	+37.5	+35		N.	9—10—8

Meteorological Journal for August, 1847.

| Barometer and Thermometer attached. || Remarks on the Weather, &c. |
Barom.	Thermo.	
30.054	+56	b. c.
30.057	+56.5	b. c.
30.051	+48.5	b. c. q. p. r.; at 5 p.m. a heavy squall and showers of rain.
29.93	+41.5	b. c. q. p. s.
30.169	+46.5	b. c.; frost last night.
30.124	+54	b. c. Ther. at 5 p.m. +62°—; all the large and deep lakes still covered with ice.
30.035	+61	b. c. q.
29.806	+54	o. p. r.
29.882	+47	b. c. q.
29.732	+43	o. r. s. s. b. c.

Figures and Letters used for denoting the state of the Weather and the force of the Wind, as recommended by Captain (now Admiral) Beaufort.

0—Calm.	b.—Blue sky.
1—Light air.	c.—Cloudy.
2—Light breeze.	d.—Drizzling rain.
3—Gentle breeze.	f.—Foggy.
4—Moderate breeze.	g.—Gloomy dark weather.
5—Fresh breeze.	h.—Hail
6—Strong breeze.	l.—Lightning
7—Moderate gale.	m.—Misty hazy atmosphere.
8—Fresh gale.	o.—Overcast.
9—Strong gale.	p.—Passing temporary showers.
10—Whole gale.	q.—Squally.
11—Storm.	r.—Rain—continued rain.
12—Hurricane.	s—Snow.
	t.—Thunder.
	u.—Ugly, threatening appearance of the weather.
	v.—Visibility of distant objects whether the sky be cloudy or not.
	w.—Wet dew.
	.—Under any letter indicates an extraordinary degree.

At the back of the original edition of *Narrative of an expedition to the shores of the Arctic Sea, in 1846 and 1847* by John Rae, the publisher, T & W Boone, included a number of notices for other books they published on expeditions and discoveries. They are reprinted here.

LIBRARY OF
AUSTRALIAN TRAVELS, &c.
PUBLISHED BY T. AND W. BOONE,
29, NEW BOND STREET.

Now ready, in 2 vols. 8vo. with numerous Plates, some coloured,

NARRATIVE OF AN EXPEDITION
INTO
CENTRAL AUSTRALIA,
BY ORDER OF HER MAJESTY'S GOVERNMENT,
DURING THE YEARS 1844, 5, 6,

With Notices of the Colony of South Australia.

BY CAPTAIN CHARLES STURT,
LATE 39TH REGT.
COLONIAL TREASURER, AND
AUTHOR OF "TWO EXPEDITIONS INTO SOUTHERN AUSTRALIA."

The character of the far interior of Australia had long been a most interesting geographical problem, many imagining the centre to be occupied by a large inland sea, others conjecturing that it was an arid desert, which opinion was further strengthened by Mr. Eyre's unsuccessful endeavour to penetrate higher than the 29th degree of latitude in his expedition during the years 1840 and 1. Captain Sturt, so appropriately denominated the "Father of Australian Discovery," in consequence of being the first traveller to explore the rivers Murray, Murrumbidgee, Bogan, and Castlereagh, volunteered to conduct a party into the interior to determine this important question. With the approbation of Lord Stanley, the Colonial Minister, he accordingly started in the year 1844, and, after a series of unparalleled privations, succeeded in reaching the centre of the Continent in a line direct north of Adelaide. The journal of this perilous Expedition gives an account of the remarkable Stony Desert, the bed of Lake Torrens, descriptions of the Natives and their villages, and the discovery of several small rivers, &c.; added to which, his observations and collections on the Natural History have since been arranged by R. Brown, Esq. and J. Gould, Esq. in the form of an Appendix.

" 'The details of this romantic and perilous Expedition are replete with interest. From the numerous and lengthened expeditions he has undertaken, and the general intelligence and scientific skill he brings to bear upon the question, we know of no recent traveller in Australia whose opinions are entitled to more weight.—The portion of the work which refers to the Colony of South Australia is particularly valuable to intending emigrants."—*Morning Herald.*

JOURNALS OF EXPEDITIONS OF DISCOVERY

IN

NORTH-WEST AND WESTERN AUSTRALIA,

DURING THE YEARS 1837, 1838, AND 1839,

Under the Authority of her Majesty's Government.

With Observations on the Agricultural and Commercial Capabilities and Prospects of several newly-explored fertile Regions, including

AUSTRALIND,

and on the Moral and Physical Condition of the Aboriginal Inhabitants, &c. &c.

By GEORGE GREY, Esq., LATE CAPTAIN 83RD REGT.

FORMERLY GOVERNOR OF SOUTH AUSTRALIA, NOW GOVERNOR OF NEW ZEALAND.

With Two large Maps by J. Arrowsmith, and numerous Illustrations, some coloured, in 2 vols. 8vo.

" It is not with the slightest hope of satisfying curiosity, or to anticipate the interest which the public in general, and geographers especially, always feel in enterprises of this nature, but merely to give such a sketch of the principal features of the expedition as may serve to direct those who are desirous of obtaining information respecting a portion of this remarkable country—hitherto only visited by Tasman, Dampier, Baudin, and King, and never before, we believe, penetrated by an European—to look forward to the detailed journals of the spirited officers who had the conduct of the expedition."
From Geographical Transactions.

A great portion of the country described in this Journal has never before been visited by any European. The Eastern coast of Short's Bay was for the first time seen and explored during the progress of these expeditions.

" We have rarely seen a more interesting book; it is full of splendid description and startling personal adventure; written in a plain, manly, unaffected style."—*Examiner.*

" It is impossible to have perused these highly interesting and important volumes without being inspired with feelings of warm admiration for the indomitable perseverance and heroical self-devotion of their gallant and enterprising author. Setting aside the vastly important results of Captain Grey's several expeditions, it is hardly possible to conceive narratives of more stirring interest than those of which his volumes are for the most part composed."—*United Service Gazette.*

" We have not read such a work of Travels for many years; it unites the interest of a romance with the permanent qualities of an historical and scientific treatise."—*Atlas.*

" We recommend our readers to the volumes of Captain Grey, assuring them they will derive both amusement and instruction from the perusal."—*Times.*

" This is a work deserving high praise. As a book of Travels it is one of the most interesting we remember to have met with."—*Westminster Review.*

" A book which should be in every lending library and book-club."
Englishman's Magazine.

" The contents of these interesting volumes will richly repay an attentive perusal."
Emigration Gazette.

"These narratives are replete with interest, and blend information and amusement in a very happy manner."—*Australian Magazine.*

Just published, in 1 vol. 8vo. with Plates and Woodcuts,

JOURNAL

OF AN

OVERLAND EXPEDITION IN AUSTRALIA,

FROM

MORETON BAY TO PORT ESSINGTON.

A distance of upwards of 3000 *miles.*

BY DR. LUDWIG LEICHHARDT.

N.B. A large 3 sheet Map of the Route by J. Arrowsmith is published, and to be had separately in a Case, price 9s.

OPINIONS OF THE PRESS.

" A work of unquestionable merit and utility, and its author's name will justly stand high upon the honourable list of able and enterprising men, whose courage, perseverance, and literary abilities have contributed so largely to our knowledge of the geography and productions of our distant southern colonies."—*Blackwood's Mag.*

" For the courage with which this lengthened and perilous journey was undertaken, the skill with which it was directed, and the perseverance with which it was performed, it is almost unrivalled in the annals of exploring enterprise. It richly deserves attention."—*Britannia.*

" The narrative in which he relates the results of this remarkable journey, and the extraordinary fatigues and privations endured by himself and his fellow travellers, is not merely valuable for its facts, but full of absorbing interest as a journal of perilous adventures."—*Atlas.*

" The volume before us comprises the narrative of one of the most remarkable enterprises ever planned by man's sagacity and executed by man's courage and endurance. To our minds there is in every point of view an inexpressible charm in such a book as this. It not merely narrates to us the opening of a new material world for human enterprise and scientific investigation, but it makes more clearly known to us the wondrous powers and capacities of human nature. We recommend it to our readers as a work scarcely less remarkable for the extraordinary enterprise recorded in it, than for the simplicity and modesty with which it is related."—*Morning Herald.*

" The result of his enterprise was thoroughly successful. It has added not a little to our existing stock of knowledge in the various departments of natural history, and has made discovery in districts before untrodden, of an almost boundless extent of fertile country."—*Examiner.*

" The most striking feature in the expedition is its successful accomplishment, which is of itself sufficient to place Dr. L. in the first rank of travellers. How much Dr. L. has added to geographical discovery can only be felt by an examination of the admirable maps which accompany the volume. These have been deduced on a large scale from the traveller's sketches by Mr. Arrowsmith, and engraved with a distinctness of execution, and a brief fulness of descriptive remark which leave nothing to be desired."
Spectator.

Lately published, in 2 vols. 8vo. cloth, with 8 Maps and Charts, and 57 Illustrations
BY COMMAND OF THE LORDS COMMISSIONERS OF THE ADMIRALTY.

DISCOVERIES IN AUSTRALIA
OF THE
VICTORIA, ADELAIDE, ALBERT, AND FITZROY RIVERS,
AND EXPEDITIONS INTO THE INTERIOR;
DURING THE
VOYAGE OF H.M.S. BEAGLE,
BETWEEN THE YEARS 1837 AND 1843: ALSO
A NARRATIVE OF THE VISITS OF H.M.S. BRITOMART,
COMMANDER OWEN STANLEY, R.N., F.R.S.
TO THE ISLANDS IN THE ARAFURA SEA.
BY CAPT. J. LORT STOKES, R.N.

" The whole narrative is so captivating, that we expect to find the work as much in demand at circulating libraries as at institutions of graver pretensions."—*Colon. Gaz.*

" We have to thank Capt. Stokes for a most valuable work, one that will place his name by the side of Vancouver, Tasman, Dampier, and Cook."—*New Quar. Review.*

" The science of Navigation owes a deep debt to Captain Stokes. The information contained in the present volumes must render them an invaluable companion to any ship performing a voyage to that part of the world."—*Foreign Quarterly Review.*

" Every part of it is full of matter, both for the general and scientific reader. With the acts of throwing the lead, taking angles, &c. lively anecdotes and pleasing ideas are constantly associated, so that we very much doubt whether any reader will lay aside the book, large as it is, without regret. In some parts you have all the breathless excitement of a voyage of discovery, and sail up new rivers, and explore new lands, while elsewhere your thoughts are directed to the tracks of commerce and political speculation. Altogether the work is a charming specimen of nautical literature, written in a pure, flexible, terse, and elegant style, and bespeaks everywhere in the author a mind endued with very high moral and intellectual qualities."—*Fraser's Mag.*

" While these volumes must prove of great value to the maritime profession, to the geographer, and to emigrants, they cannot fail to be perused with interest by readers in general."—*Athenæum.*

" We cannot, in noticing these two ably written and interesting volumes, insist too strongly upon their importance alike to the mariner, the geographer, and the general reader. The author is a man of considerable merit, a shrewd observer of men and things, and who was fitted by nature and inclination to conduct these researches into the vast unknown continent whither he proceeded with enterprise and spirit. These volumes contain a fund of interesting matter, and we warmly recommend this valuable addition to our literary and scientific stores to the attention of the public."
Sentinel.

" The contents of these volumes, rich, varied and full of interest, will be their best recommendation. For scientific accuracy, they will be highly valued by the geographer and navigator, while they will be read for mere amusement by the public at large."
Sunday Times.

THE EASTERN ARCHIPELAGO.

By Permission of the Lords Commissioners of the Admiralty.
Now ready, in 2 vols. 8vo. with numerous Maps, Plates, and Woodcuts,

NARRATIVE
OF THE
SURVEYING VOYAGE OF H.M.S. FLY,
UNDER THE COMMAND OF
CAPTAIN BLACKWOOD, R.N.
IN TORRES STRAIT, NEW GUINEA, AND OTHER ISLANDS
IN THE EASTERN ARCHIPELAGO;
TOGETHER
WITH AN EXCURSION INTO THE INTERIOR
OF THE
EASTERN PART OF THE ISLAND OF JAVA,
DURING THE YEARS 1842 TO 1846.
BY J. BEETE JUKES, M.A.
NATURALIST TO THE EXPEDITION.

OPINIONS OF THE PRESS.

" We must congratulate Mr. Jukes on the value of his publication. Scientific without being abstruse, and picturesque without being extravagant, he has made his volumes a striking and graceful addition to our knowledge of countries highly interesting in themselves, and assuming hourly importance in the eyes of the people of England."—*Blackwood's Magazine.*

" To transcribe the title-page of this book is sufficient to attract public curiosity towards it—to peruse the book itself is to be rewarded with the knowledge of a mass of information in which complete confidence can be reposed, for, from the first page to the last, it is apparent that the main object with Mr. Jukes is to tell all that he knows and believes to be true, rather than to win favour from his readers by his manner of telling it. There is not a pretty phrase, an exaggeration, nor an invention in the two volumes of Mr. Jukes; all is plain unadorned fact, and because it is so, is deserving, not merely of perusal, but of study. Such are the recommendations of Mr. Jukes' pages to the public, and all who desire to see truth united with novelty will peruse them."—*Morning Herald.*

" Mr. Jukes has been most judicious in his selection of topics whereon to dwell in his narrative, and he describes with great vivacity and picturesque power. There is much novelty and freshness in his book, and much valuable information."
Daily News.

" There are very few pages in the work which are not readable and entertaining."
Morning Post.

" Captain Blackwood having waived his right of authorship, the narrative of the voyage has been undertaken by Mr. Jukes, favourably known by an agreeable and informing book on Newfoundland, nor will the present work detract from his reputation. The narrative is well planned, pleasantly written, and full of matter."
Spectator.

" A great deal was seen, and Geography, Topography, Geology, Natural History, Ethnology, Philology, and Commerce may all be benefited by the work before us."
Literary Gazette.

" Mr. Jukes has performed his portion of the work with great ability, sparing no pains in the working up of his abundant material, so as to make it a book of science, as well as a book of amusement."—*Critic.*

" Although a professed man of science, he has described what he saw in a lucid and untechnical manner, so that his work will be found interesting to the ordinary reader, while it is equally valuable to the scientific. The amount of information conveyed is very great."—*Midland Herald.*

In 2 vols. 8vo. with Maps and numerous Plates,

JOURNALS OF EXPEDITIONS OF DISCOVERY

INTO

CENTRAL AUSTRALIA,

AND

OVERLAND FROM ADELAIDE TO KING GEORGE'S SOUND,

IN THE YEARS 1840-1;

Sent by the Colonists of South Australia,

WITH THE SANCTION AND SUPPORT OF THE GOVERNMENT:

INCLUDING

An Account of the Manners and Customs of the Aborigines, and the state of their relations with Europeans.

BY EDWARD JOHN EYRE,

RESIDENT MAGISTRATE, MURRAY RIVER, NOW LIEUT.-GOVERNOR OF NEW ZEALAND.

*** The Founder's Medal of the Royal Geographical Society was awarded to Mr. Eyre for the discovery of Lake Torrens, and explorations of far greater extent in Australia than any other traveller, a large portion never having been previously traversed by civilized man.

"His narrative of what he did and overcame, is more like the stirring stories of Park and Bruce than the tame and bookish diffuseness of modern travellers. Nothing short of a perusal of the volumes can enable our readers to appreciate this book."—*Spectator.*

"We might easily extract much more from Mr. Eyre's volumes of interest to the reader, but our limits circumscribe us. We therefore bid farewell to them, with the recommendation to the public, not to overlook a work which, though it records the failure of a great enterprize, is yet full of matter, which proclaims it of value."
Atlas.

"Mr. Eyre writes with the plain unaffected earnestness of the best of the old travellers."—*Examiner.*

"An intensely interesting book."—*Tablet.*

"We must now close these interesting volumes, not, however, without expressing our high approval both of the matter they contain, and of the manner of their compilation. We rise from the perusal of them with a feeling similar to that which follows the enjoyment of a pleasant work of fiction."—*Critic.*

In 1 vol. 8vo. cloth, with large Map by Arrowsmith, and numerous Illustrations,

SOUTH AUSTRALIA AND ITS MINES,

With an *Historical Sketch of the Colony, under its several Administrations, to the Period of Captain Grey's Departure.*

By FRANCIS DUTTON.

" The best work which has yet issued from the press, descriptive of the resources and management of this thriving colony."—*Mining Journal.*

" We have here a well-timed book. South Australia and Its Mines are now objects of great interest; and Mr. Dutton's plain, unadorned recital, contains just what the intending emigrant, or the mercantile inquirer, will rejoice at having placed within his reach."—*Colonial Gazette.*

COLONIZATION;
PARTICULARLY IN SOUTHERN AUSTRALIA,

WITH SOME REMARKS ON SMALL FARMS AND OVER POPULATION.

BY MAJOR-GENERAL SIR CHARLES JAMES NAPIER, K.C.B.

Author of " The Colonies; particularly the Ionian Islands "

In 1 vol. 8vo. price 7s. boards.

" We earnestly recommend the book to all who feel an interest in the welfare of the people."—*Sun.*

In 1 vol. post 8vo. price 5s. 6d.

HINTS FOR AUSTRALIAN EMIGRANTS,

WITH

ENGRAVINGS AND EXPLANATORY DESCRIPTIONS
OF THE WATER RAISING WHEELS,

AND MODES OF IRRIGATING LAND IN SYRIA, EGYPT, SOUTH AMERICA, ETC.

BY PETER CUNNINGHAM,

SURGEON, R.N.

Author of " *Two Years in New South Wales,*" &c.

" The mere name of Mr. Cunningham affords an ample guarantee for the value of any work to which it may be prefixed; and, " to all whom it may concern," we can confidently recommend this remarkably neat little volume as replete with practical information. Its numerous illustrative engravings in wood are executed in a very superior style."—*Naval and Military Gazette, October 23rd,* 1841.

In 1 vol. 8vo. Map and Plates, cloth, price 12s.

AUSTRALIA,
FROM PORT MACQUARIE TO MORETON BAY,
WITH

Descriptions of the Natives, their Manners and Customs, the Geology, Natural Productions, Fertility, and Resources of that Region.

First explored and surveyed by order of the Colonial Government.

BY CLEMENT HODGKINSON.

"The work before our consideration contains certain details connected with the portion of Australia, described in it, which will prove of first-rate importance to the colonist and emigrant, since they are evidently derived from practical experience. Throughout this unpretending little work we trace great honesty of purpose, and a disposition to state no more than the bare facts as they presented themselves."

New Quarterly Review.

Just published, in 2 vols. 8vo. with a large Map,

AN HISTORICAL, POLITICAL, AND STATISTICAL ACCOUNT

OF THE

ISLAND OF CEYLON.

By CHARLES PRIDHAM, Esq.

AUTHOR OF

"THE MAURITIUS AND ITS DEPENDENCIES."

"——— All these events will be found fully set forth in the volumes under notice, which are certainly far superior as a history of Ceylon to any other that has yet appeared. The reader will also find in these pages curious and original information respecting the habits, manners and customs of the Cingalese, which he may look for in vain in similar publications. Every portion of this valuable work teems with information of a precise and important character."—*Observer.*

"Those who seek information on the subject of Ceylon, will find his book a great storehouse of facts."—*Economist.*

JUST PUBLISHED,

A SERIES OF TEN COLOURED VIEWS,

TAKEN DURING THE ARCTIC EXPEDITION OF HER MAJESTY'S SHIPS

ENTERPRISE and INVESTIGATOR,

UNDER THE COMMAND OF

Captn. SIR JAMES C. ROSS, Kt. F. R. S.,

IN SEARCH OF

Captn. SIR JOHN FRANKLIN, Kt. K. C. H.,

Drawn by Lieut. W. H. BROWNE, R.N.

LATE OF H. M. S. ENTERPRISE,

With a Summary of the Arctic Expeditions in search of Sir John Franklin.

Dedicated, by Special Permission,

TO THE LORDS COMMISSIONERS OF THE ADMIRALTY.

Price, in a Cover .... **16s.**
Handsomely bound .... **21s.**

LONDON.
ACKERMANN & CO., 96, STRAND,

By Appointment to H. M. the Queen, H. R. H. Prince Albert,
H. R. H. the Duchess of Kent, and the Royal Family.

OPINIONS OF THE PRESS.

———The ex+reme interest evinced by the public would be likely to secure a welcome for these views if their execution had been less felicitous than it is. The Party arriving at the Southern Depôt is fearfully grand.———

ATHENÆUM.

———Such are these ten extraordinary views; revealing scenes which are enough to appal the stoutest hearts. We seem to ask of these mountains of thick-ribbed ice " are our countrymen hidden from us by your fantastic forms?" &c.———

LITERARY GAZETTE.

———We do not remember ever being so powerfully impressed with the sublimity of portfolio drawings as with some of these views of the *icy Polar Regions* of the trackless North.———

UNITED SERVICE GAZETTE.

———We do not speak of it as a work of art merely, but of the evident truth of delineation, of local colouring, and atmospheric effects.———

GLOBE.

———This is a work which will no doubt meet with general patronage—giving a vivid idea of the frozen regions.———

BELL'S LIFE.

———Ten of the most interesting views which scenery can furnish.———

ATLAS.

———Perhaps the most attractive, as well as most effective, is *Noon in Mid-Winter*, and conveys the most solemn notions of the *Polar Regions*. This portfolio is the novelty of the season.———

CRITIC.

PART THREE

JOHN RAE CONTENDS WITH CHARLES DICKENS

In 1854, Rae returned to England with news of the rather shocking fate of the Franklin expedition. After his brief report to the Admiralty, which he never intended for public consumption, was reprinted in *The London Times*, Charles Dickens set out to publicly discredit Rae's findings. Over the course of December 1854, the two men engaged in a battle of words in the pages of Dickens's own publication, *Household Words*.

"Part Three" comprises various pieces published after John Rae returned to London in 1854. They include Rae's original report to the Admiralty and his letter to the editor, as they were published in *The Times* on October 23, 1854. As well, we have here the two-part rebuttal Charles Dickens offered in *Household Words*, on December 2 and 9, 1854, and John Rae's response, which appeared in that same journal, on December 23 and 30, 1854. As the final piece in this public conversation, we include John Rae's report to the Hudson's Bay Company, originally written at York Factory on September 1, 1854, and published by Charles Dickens in *Household Words* on May 1, 1855.

THE ARCTIC EXPEDITION

John Rae's report to the Secretary of the Admiralty and his letter to the editor of *The London Times*, originally published October 23, 1854.

Note: John Rae did not anticipate that his report to the Admiralty would appear in *The Times*. He submitted only the short, detail-free letter that the newspaper printed beneath the report.

Intelligence which may be fairly considered decisive has at last reached this country of the sad fate of Sir John Franklin and his brave companions.

Dr. Rae, whose previous exploits as a Arctic traveller have already so highly distinguished him, landed at Deal yesterday, and immediately proceeded to the Admiralty, and laid before Sir James Graham the melancholy evidence on which his report is founded.

Dr. Rae was not employed in searching for Sir John Franklin, but in completing his survey of the coast of Boothia. He justly thought, however, that the information he had obtained greatly outweighed the importance of his survey, and he has hurried home to satisfy the public anxiety as to the fate of the long-lost expedition, and to prevent the risk of any more lives in a fruitless search. It would seem from his description of the place in which the bodies were found that both Sir James Ross and Captain Bellot must have been within a few miles of the spot to which our unfortunate countrymen had struggled on in their desperate march. A few of the unfortunate men must, he thinks, have survived until the arrival of the wildfowl about the end of May, 1850, as shots were heard and fresh bones and feathers of geese were noticed near the scene of the sad event.

We subjoin Dr. Rae's report to the Admiralty, and a letter with which he has favoured us:—

The following is Dr. Rae's report to the Secretary of the Admiralty:—

Repulse Bay, July 29, 1854.

Sir,—I have the honour to mention, for the information of my Lords Commissioners of the Admiralty, that, during my journey over the ice and snows this spring, with the view of completing the survey of the west shore of Boothia, I met with Esquimaux in Pelly Bay, from one of whom I learnt that a party of "white men" (Kabloonans) had perished from want of food some distance to the westward, and not far beyond a large river containing many falls and rapids. Subsequently, further particulars were received and a number of articles purchased, which places the fate of a portion, if not of all, of the then survivors of Sir John Franklin's long-lost party beyond a doubt—a fate as terrible as the imagination can conceive.

The substance of the information obtained at various times and from various sources was as follows:—

In the spring, four winters past (spring, 1850), a party of "white men," amounting to about 40, were seen travelling southward over the ice and dragging a boat with them by some Esquimaux, who were killing seals near the north shore of King William Land, which is a large island. None of the party could speak the Esquimaux language intelligibly, but by signs the natives were made to understand that their ship, or ships, had been crushed by ice, and that they were now going to where they expected to find deer to shoot. From the appearance of the men, all of whom except one officer looked thin, they were then supposed to be getting short of provisions, and they purchased a small seal from the natives. At a later date the same season, but previously to the breaking up of the ice, the bodies of some 30 persons were discovered on the continent, and five on an island near it, about a long day's journey to the N.W. of a large stream, which can be no other than Back's Great Fish River (named by the Esquimaux Oot-ko-hi-ca-lik), as its description and that of the low shore in the neighbourhood of Point Ogle and Montreal Island agree exactly with that of Sir George Back. Some of the bodies had been buried (probably those of the first victims of famine); some were in a tent or tents; others under the boat, which had been turned over to form a shelter, and several lay scattered about in different directions. Of those found on the island one was supposed to have been an officer, as he had a telescope strapped over his shoulders and his double-barrelled gun lay underneath him.

From the mutilated state of many of the corpses and the contents of the kettles, it is evident that our wretched countrymen had been driven to the last resource—cannibalism—as a means of prolonging existence.

There appeared to have been an abundant stock of ammunition, as the powder was emptied in a heap on the ground by the natives out of the kegs or cases containing it; and a quantity of ball and shot was found below high-water mark, having probably been left on the ice close to the beach. There must have been a number of watches, compasses, telescopes, guns (several double-barrelled), &c., all of which appear to have been broken up, as I saw pieces of these different articles with the Esquimaux, and, together with some silver spoons and forks, purchased as many as I could get. A list of the most important of these I enclose, with a rough sketch of the crests and initials on the forks and spoons. The articles themselves shall be handed over to the Secretary of the Hon. Hudson's Bay Company on my arrival in London.

None of the Esquimaux with whom I conversed had seen the "whites," nor had they ever been at the place where the bodies were found, but had their information from those who had been there and who had seen the party when travelling.

I offer no apology for taking the liberty of addressing you, as I do so from a belief that their Lordships would be desirous of being put in possession at as early a date as possible of any tidings, however meagre and unexpectedly obtained, regarding this painfully interesting subject.

I may add that, by means of our guns and nets, we obtained an ample supply of provisions last autumn, and my small party passed the winter in snow houses in comparative comfort, the skins of the deer shot affording abundant warm clothing and bedding. My spring journey was a failure in consequence of an accumulation of obstacles, several of which my former experience in Arctic travelling had not taught me to expect. I have, &c.,

JOHN RAE, C.F.,
Commanding Hudson's Bay Company's Arctic Expedition.

List of Articles purchased from the Esquimaux, said to have been obtained at the place where the bodies of the persons reported to have died of famine were found, viz.—

1 silver table fork—crest, an animal's head with wings, extended above; 3 silver table forks—crest, a bird with wings extended; 1 silver table spoon—crest, with initials "F.R.M.C." (Captain Crozier, Terror); 1 silver table spoon and 1 fork—crest, bird with laurel branch in mouth, motto, "*Spero meliora*;" 1 silver table spoon, 1 tea spoon, and 1 dessert fork—crest, a fish's head looking upwards, with laurel branches on each side; 1 silver table fork—initials, "H.D.S.G." (Harry D.S. Goodsir, assistant-surgeon, Erebus); 1 silver table fork—initials, "A. McD." (Alexander McDonald, assistant-surgeon, Terror); 1 silver table fork—initials, "G.A.M." (Gillies A. Macbean, second-master, Terror); 1 silver table fork—initials, "J.T."; 1 silver dessert spoon—initials, "J.S.P." (John S. Peddie, surgeon, Erebus); 1 round silver plate, engraved, "Sir John Franklin, K.C.B."; a star or order, with motto, "*Nec aspera terrent*, G.R. III., MDCCCXV."

Also a number of other articles with no marks by which they could be recognized, but which will be handed over with those abovenamed to the Secretary of the Hon. Hudson's Bay Company.

JOHN RAE, C.F.
Repulse Bay, July, 1854.

TO THE EDITOR OF THE TIMES.

Sir,—As any information, however meagre, regarding Sir John Franklin and his party must be of deep interest to every one, I take the earliest opportunity of communicating the following particulars:—

During my journey from Repulse Bay this spring over the ice, with the view of completing the survey of the west coast of Boothia, I then and subsequently obtained information, and purchased articles of the natives, which prove beyond a doubt that a portion (if not all) of the then survivors of Sir John Franklin's long-lost and ill-fated party perished of starvation in the spring of 1850, on the coast of America, a short distance west of a large stream, which, by the description given of it, can be no other than Back's Fish River.

Among the articles purchased (all of which are now in my possession), which the Esquimaux found where the corpses of the "white men" were discovered, are a small silver plate, with "Sir John Franklin, K.C.B.," engraved upon it, several silver spoons and forks, with initials of the following officers—viz., Captain Crozier, Lieutenant G. Gore, Assistant-Surgeon A. McDonald, Assistant-Surgeon J.S. Peddie, and Second-Master G.A. McBean.

Further particulars on this melancholy subject will appear in my report to the Hon. Hudson's Bay Company.

I may add that my small party wintered in snow houses comfortably enough at Repulse Bay, after some very hard work in the autumn in laying up a supply of venison and fuel. We returned to York Factory all well on the 30th of August, but without having completed the contemplated survey.

I have the honour to be, Sir,
Your very obedient servant,
JOHN RAE, M.D.,
Commanding Hudson's Bay Company's Arctic Expedition.
Ship Prince of Wales, in the English Channel, Oct. 20.

P.S. We have had a very rough passage of 31 days from York Factory, saw a great quantity of ice in Hudson's Straits, and in a storm in the Atlantic split three or four of the principal sails, and very nearly had the lifeboat washed away. I fear the season has been a most trying one for the ships in the far north.

Dr. Rae adds, that from what he could learn there is no reason to suspect that any violence had been offered to the sufferers by the natives. It seems but too evident that they had perished from hunger, aggravated by the extreme severity of the climate. Some of the corpses had been sadly mutilated, and had been stripped by those who had the misery to survive them, and who were found wrapped in two or three suits of clothes. The articles brought home by Dr. Rae had all been worn as ornaments by the Esquimaux, the coins being pierced with holes, so as to be suspended as medals. A large number of books were also found, but these not being valued by the natives had either been destroyed or neglected. Dr. Rae has no doubt, from the careful habits of these people, that almost every article which the unhappy sufferers had preserved could be recovered, but he thought it better to come home direct with the intelligence he had obtained than to run the risk of having to spend another winter in the snow.

THE LOST ARCTIC VOYAGERS

Charles Dickens's rebuttal of claims made in Rae's report to the Admiralty, originally published in *Household Words*, December 2, 1854.

Dr. Rae may be considered to have established, by the mute but solemn testimony of the relics he has brought home, that Sir John Franklin and his party are no more. But, there is one passage in his melancholy report, some examination into the probabilities and improbabilities of which, we hope will tend to the consolation of those who take the nearest and dearest interest in the fate of that unfortunate expedition, by leading to the conclusion that there is no reason whatever to believe, that any of its members prolonged their existence by the dreadful expedient of eating the bodies of their dead companions. Quite apart from the very loose and unreliable nature of the Esquimaux representations (on which it would be necessary to receive with great caution, even the commonest and most natural occurrence), we believe we shall show, that close analogy and the mass of experience are decidedly against the reception of any such statement, and that it is in the highest degree improbable that such men as the officers and crews of the two lost ships would, or could, in any extremity of hunger, alleviate the pains of starvation by this horrible means.

Before proceeding to the discussion, we will premise that we find no fault with Dr. Rae, and that we thoroughly acquit him of any trace of blame. He has himself openly explained, that his duty demanded that he should make a faithful report, to the Hudson's Bay Company or the

Admiralty, of every circumstance stated to him; that he did so, as he was bound to do, without any reservation; and that his report was made public by the Admiralty: not by him. It is quite clear that if it were an ill-considered proceeding to disseminate this painful idea on the worst of evidence, Dr. Rae is not responsible for it. It is not material to the question that Dr. Rae believes in the alleged cannibalism; be does so, merely "on the substance of information obtained at various times and various sources," which is before us all. At the same time, we will most readily concede that he has all the rights to defend his opinion which his high reputation as a skilful and intrepid traveller of great experience in the Arctic Regions—combined with his manly, conscientious, and modest personal character—can possibly invest him with. Of the propriety of his immediate return to England with the intelligence he had got together, we are fully convinced. As a man of sense and humanity, he perceived that the first and greatest account to which it could be turned, was, the prevention of the useless hazard of valuable lives; and no one could better know in how much hazard all lives are placed that follow Franklin's track, than he who had made eight visits to the Arctic shores. With these remarks we can release Dr. Rae from this inquiry, proud of him as an Englishman, and happy in his safe return home to well-earned rest.

The following is the passage in the report to which we invite attention: "Some of the bodies had been buried (probably those of the first victims of famine); some were in a tent or tents; others under the boat, which had been turned over to form a shelter; and several lay scattered about in different directions. Of those found on the island, one was supposed to have been an officer; as he had a telescope, strapped over his shoulders, and his double-barrelled gun lay underneath him. From the mutilated state of many of the corpses and the contents of the kettles, it is evident that our wretched countrymen had been driven to the last resource—cannibalism—as a means of prolonging existence.... None of the Esquimaux with whom I conversed had seen the 'whites,' nor had they ever been at the place where the bodies were found, but had their information from those who had been there, and who had seen the party when travelling."

We have stated our belief that the extreme improbability of this inference as to the last resource, can be rested, first on close analogy, and secondly, on broad general grounds, quite apart from the improbabilities and incoherencies of the Esquimaux evidence: which is itself given, at the very best, at second-hand. More than this, we presume it to have been given at second-hand through an interpreter; and he was, in all probability, imperfectly acquainted with the language he translated to the white man. We believe that few (if any) Esquimaux tribes speak one common dialect; and Franklin's own experience of his interpreters in his former voyage was, that they and the Esquimaux they encountered understood each other "tolerably"—an expression which he frequently uses in his book, with the evident intention of showing that their communication was not altogether satisfactory. But, even making the very large admission that Dr. Rae's interpreter perfectly understood what he was told, there yet remains the question whether he could render it into language of corresponding weight and value. We recommend any reader who does not perceive the difficulty of doing so and the skill required, even when a copious and elegant European language is in question, to turn to the accounts of the trial of Queen Caroline, and to observe the constant discussions that arose—sometimes, very important—in reference to the worth in English, of words used by the Italian witnesses. There still remains another consideration, and a grave one, which is, that ninety-nine interpreters out of a hundred, whether savage, half-savage, or wholly civilised, interpreting to a person of superior station and attainments, will be under a strong temptation to exaggerate. This temptation will always be strongest, precisely where the person interpreted to is seen to be the most excited and impressed by what he hears; for, in proportion as he is moved, the interpreter's importance is increased. We have ourself had an opportunity of inquiring whether any part of this awful information, the unsatisfactory result of "various times and various sources," was conveyed by gestures. It was so, and the gesture described to us as often repeated—that of the informant setting his mouth to his own arm—would quite as well describe a man having opened one of his veins, and drunk of the stream that flowed from it. If it be inferred that the officer who lay upon his double-barrelled gun, defended his life to the last against ravenous seamen, under the boat or elsewhere, and that he died in so doing, how came his body to be found? That was not eaten, or even mutilated, according to the description. Neither were the bodies, buried in the frozen earth, disturbed; and is it not likely that if any bodies were resorted to as food, those the most removed from recent life and companionship would have been the first? Was there any fuel in that desolate place for cooking "the contents of the kettles"? If none, would the little flame of the spirit-lamp the travellers *may have had* with them, have sufficed for such a purpose? If not, would the kettles have been defiled for that purpose at all? "Some of the corpses," Dr. Rae adds, in a letter to the Times, "had been sadly mutilated, and had been stripped by those who had the misery to survive them, and who were found wrapped in two or three suits of clothes." Had there been no bears thereabout, to mutilate those bodies; no wolves, no foxes? Most probably the scurvy, known to be the dreadfullest scourge of Europeans in those latitudes, broke out among the party. Virulent as it would inevitably be under such circumstances, it would of itself cause dreadful disfigurement—woeful mutilation—but, more than that, it would not only soon annihilate the desire to eat (especially to eat flesh of any kind), but would annihilate the power. Lastly, no man can, with any show of reason, undertake to affirm that this sad remnant of Franklin's gallant band were not set upon and slain by the Esquimaux themselves. It is impossible to form an estimate of the character of any race of savages, from their deferential behaviour to the white man while he is strong. The mistake has been made again and again; and the moment the white man has appeared in the new aspect of being weaker than the savage, the savage has changed and sprung upon him. There are pious persons who, in their practice, with a strange inconsistency, claim for every child born to civilisation all innate depravity, and for every savage born to the woods and wilds all innate virtue. We believe every savage to be in his heart covetous, treacherous, and cruel; and we have yet to learn what knowledge the white man—lost, houseless, shipless, apparently forgotten by his race, plainly famine-stricken, weak, frozen, helpless, and dying—has of the gentleness of Esquimaux nature.

Leaving, as we purposed, this part of the subject with a glance, let us put a supposititious case.

If a little band of British naval officers, educated and trained exactly like the officers of this ill-fated expedition, had, on a former occasion, in command of a party of men vastly inferior to the crews of these two ships, penetrated to the same regions, and been exposed to the rigours of the same climate; if they had undergone such fatigue, exposure, and disaster, that scarcely power remained to them to crawl, and they tottered and fell many times in a journey of a few yards; if they could not bear the contemplation of their "filth and wretchedness, each other's emaciated figures, ghastly countenances, dilated eyeballs, and sepulchral voices"; if they had eaten their shoes, such outer clothes as they could part with and not perish of cold, the scraps of acrid marrow yet remaining in the dried and whitened spines of dead wolves; if they had wasted away to skeletons, on such fare, and on bits of putrid skin, and bits of hide, and the covers of guns, and pounded bones; if they had passed through all the pangs of famine, had reached that point of starvation where there is little or no pain left, and had descended so far into the valley of the shadow of Death, that they lay down side by side, calmly and even cheerfully awaiting their release from this world; if they had suffered such dire extremity, and yet lay where the bodies of their dead companions lay unburied, within a few paces of them; and yet never dreamed at the last gasp of resorting to this said "last resource;" would it not be strong presumptive evidence against an incoherent Esquimaux story, collected at "various times" as it wandered from "various sources"? But, if the leader of that party were the leader of this very party too; if Franklin himself had undergone those dreadful trials, and had been restored to health and strength, and had been—not for days and months alone, but years—the Chief of this very expedition, infusing into it, as such a man necessarily must, the force of his character and discipline, patience and fortitude; would there not be a still greater and stronger moral improbability to set against the wild tales of a herd of savages?

Now, this *was* Franklin's case. He had passed through the ordeal we have described. He was the Chief of that expedition, and he was the Chief of this. In this, he commanded a body of picked English seamen of the first class; in that, he and his three officers had but one English seaman to rely on; the rest of the men being Canadian voyagers and Indians. His Narrative of a Journey to the Shores of the Polar Sea in 1819–22, is one of the most explicit and enthralling in the whole literature of Voyage and Travel. The facts are acted and suffered before the reader's eyes, in the descriptions of Franklin, Richardson, and Back: three of the greatest names in the history of heroic endurance.

See how they gradually sink into the depths of misery.

"I was reduced," says Franklin, long before the worst came, "almost to skin and bone, and, like the rest of the party, suffered from degrees of cold that would have been disregarded whilst in health and vigour." "I set out with the intention of going to Saint Germain, to hasten his operations (making a canoe), but though he was only three quarters of a mile distant, I spent three hours in a vain attempt to reach him, my strength being unequal to the labour of wading through the deep snow; and I returned quite exhausted, and much shaken by the numerous falls I had got. My associates were all in the same debilitated state. The voyagers were somewhat stronger than ourselves, but more indisposed to exertion, on account of their despondency. The sensation of hunger was no longer felt by any of us, yet we were scarcely able to converse upon any other subject than the pleasures of eating." "We had a small quantity of this weed (tripe de roche, and always the cause of miserable illness to some of them) in the evening, and the rest of our supper was made up of scraps of roasted leather. The distance walked to-day was six miles." "Previous to setting out, the whole party ate the remains of their old shoes, and whatever scraps of leather they had, to strengthen their stomachs for the fatigue of the day's journey." "Not being able to find any tripe de roche, we drank an infusion of the Labrador tea-plant, and ate a few morsels of burnt leather for supper." "We were unable to raise the tent, and found its weight too great to carry it on; we therefore cut it up, and took a part of the canvass for a cover." Thus growing weaker and weaker every day, they reached, at last, Fort Enterprise, a lonely and desolate hut, where Richardson—then

Dr. Richardson, now Sir John—and Hepburn, the English seaman, from whom they had been parted, rejoined them. "We were all shocked at beholding the emaciated countenances of the Doctor and Hepburn, as they strongly evidenced their extremely debilitated state. The alteration in our appearance was equally distressing to them, for, since the swellings had subsided, we were little more than skin and bone. The Doctor particularly remarked the sepulchral tone of our voices, which he requested us to make more cheerful, if possible, quite unconscious that his own partook of the same key." "In the afternoon Peltier was so much exhausted, that he sat up with difficulty, and looked piteously; at length he slided from his stool upon the bed, as we supposed to sleep, and in this composed state he remained upwards of two hours without our apprehending any danger. We were then alarmed by hearing a rattling in his throat, and on the Doctor's examining him he was found to be speechless. He died in the course of the night. Semandré sat up the greater part of the day, and even assisted in pounding some bones; but, on witnessing the melancholy state of Peltier, he became very low, and began to complain of cold, and stiffness of the joints. Being unable to keep up a sufficient fire to warm him, we laid him down, and covered him with several blankets. He did not, however, appear to get better, and I deeply lament to add, he also died before daylight. We removed the bodies of the deceased into the opposite part of the house, but our united strength was inadequate to the task of interring them, or even carrying them down to the river." "The severe shock occasioned by the sudden dissolution of our two companions, rendered us very melancholy. Adam (one of the interpreters) became low and despondent; a change which we lamented the more, as we perceived he had been gaining strength and spirits for the two preceding days. I was particularly distressed by the thought that the labour of collecting wood must now devolve upon Dr. Richardson and Hepburn, and that my debility would disable me from affording them any material assistance; indeed both of them most kindly urged me not to make the attempt. I found it necessary, in their absence, to remain constantly near Adam and to converse with him, in order to prevent his reflecting on our condition, and to keep up his spirits as far as possible. I also lay by his side at night." "The Doctor and Hepburn were getting much weaker, and the limbs of the latter were now greatly swelled. They came into the house frequently in the course of the day to rest themselves, and when once seated were unable to rise without the help of one another, or of a stick. Adam was for the most part in the same low state as yesterday, but sometimes he surprised us by getting up and walking with an appearance of increased strength. His looks were now wild and ghastly, and his conversation was often incoherent." "I may here remark, that owing to our loss of flesh, the hardness of the floor, from which we were only protected by a blanket, produced soreness over the body, and especially those parts on which the weight rested in lying; yet to turn ourselves for relief was a matter of toil and difficulty. However, during this period, and indeed all along after the acute pains of hunger, which lasted but a short time, had subsided, we generally enjoyed the comfort of a few hours' sleep. The dreams which for the most part but not always accompanied it, were usually (though not invariably) of a pleasant character, being very often about the enjoyments of feasting. In the daytime, we fell into the practice of conversing on common and light subjects, although we sometimes discoursed, with seriousness and earnestness, on topics connected with religion. We generally avoided speaking, directly, of our present sufferings, or even of the prospect of relief. I observed, that in proportion as our strength decayed, our minds exhibited symptoms of weakness, evinced by a kind of unreasonable pettishness with each other. Each of us thought the other weaker in intellect than himself, and more in need of advice and assistance. So trifling a circumstance as a change of place, recommended by one as being warmer and more comfortable, and refused by the other from a dread of motion, frequently called forth fretful expressions, which were no sooner uttered than atoned for, to be repeated, perhaps, in the course of a few minutes. The same thing often occurred when we endeavoured to assist each other in carrying wood to the fire; none of us were willing to receive assistance, although the task was disproportioned to our strength. On one of these occasions, Hepburn was so convinced of this waywardness, that he exclaimed, 'Dear me, if we are spared to return to England, I wonder if we shall recover our understandings!'"

Surely it must be comforting to the relatives and friends of Franklin and his brave companions in later dangers, now at rest, to reflect upon this manly and touching narrative; to consider that at the time it so affectingly describes, and all the weaknesses of which it so truthfully depicts, the bodies of the dead lay within reach, preserved by the cold, but unmutilated; and to know it for an established truth, that the sufferers had passed the bitterness of hunger and were then dying passively.

They knew the end they were approaching very well, as Franklin's account of the arrival of their deliverance next day, shows. "Adam had passed a restless night, being disquieted by gloomy apprehensions of approaching death, which we tried in vain to dispel. He was so low in the morning as to be scarcely able to speak. I remained in bed by his side, to cheer him as much as possible. The Doctor and Hepburn went to cut wood. They had hardly begun their labour, when they were amazed at hearing the report of a musket. They could scarcely believe that there was really anyone near, until they heard a shout, and immediately espied three Indians close to the house. Adam and I heard the latter noise, and I was fearful that a part of the house had fallen upon one of my companions; a disaster which had in fact been thought not unlikely. My alarm was only momentary. Dr. Richardson came in to communicate the joyful intelligence that relief had arrived. He and myself immediately addressed thanksgiving to the throne of mercy for this deliverance, but poor Adam was in so low a state that he could scarcely comprehend the information. When the Indians entered, he attempted to rise, but sank down again. But for this seasonable interposition of Providence, his existence must have terminated in a few hours, and that of the rest probably in not many days."

But, in the preceding trials and privations of that expedition, there *was* one man, Michel, an Iroquois hunter, who *did* conceive the horrible idea of subsisting on the bodies of the stragglers, if not of even murdering the weakest with the express design of eating them—which is pretty certain.

This man planned and executed his wolfish devices at a time when Sir John Richardson and Hepburn were afoot with him every day; when, though their sufferings were very great, they had not fallen into the weakened state of mind we have just read of; and when the mere difference between his bodily robustness and the emaciation of the rest of the party—to say nothing of his mysterious absences and returns—might have engendered suspicion. Yet, so far off was the unnatural thought of cannibalism from their minds, and from that of Mr. Hood, another officer who accompanied them—though they were all then suffering the pangs of hunger, and were sinking every hour—that no suspicion of the truth dawned upon one of them, until the same hunter shot Mr. Hood dead as he sat by a fire. It was after the commission of that crime, when he had become an object of horror and distrust, and seemed to be going savagely mad, that circumstances began to piece themselves together in the minds of the two survivors, suggesting a guilt so monstrously unlikely to both of them that it had never flashed upon the thoughts of either until they knew the wretch to be a murderer. To be rid of his presence, and freed from the danger they at length perceived it to be fraught with, Sir John Richardson, nobly assuming the responsibility he would not allow a man of commoner station to bear, shot this devil through the head—to the infinite joy of all the generations of readers who will honour him in his admirable narrative of that transaction.

The words in which Sir John Richardson mentions this Michel, after the earth is rid of him, are extremely important to our purpose, as almost describing the broad general ground towards which we now approach. "His principles, unsupported by a belief in the divine truths of Christianity, were unable to withstand the pressure of severe distress. His countrymen, the Iroquois, are generally Christians, but he was totally uninstructed, and ignorant of the duties inculcated by Christianity; and from his long residence in the Indian country, seems to have imbibed, or retained, the rules of conduct which the southern Indians prescribe to themselves."

Heaven forbid that we, sheltered and fed, and considering this question at our own warm hearth, should audaciously set limits to any extremity of desperate distress! It is in reverence for the brave and enterprising, in admiration for the great spirits who can endure even unto the end, in love for their names, and in tenderness for their memory, that we think of the specks, once ardent men, "scattered about in different directions" on the waste of ice and snow, and plead for their lightest ashes. Our last claim in their behalf and honour, against the vague babble of savages, is, that the instances in which this "last resource" so easily received, has been permitted to interpose between life and death, are few and exceptional; whereas the instances in which the sufferings of hunger have been borne until the pain was past, are very many. Also, and as the citadel of the position, that the better educated the man, the better disciplined the habits, the more reflective and religious the tone of thought, the more gigantically improbable the "last resource" becomes.

Beseeching the reader always to bear in mind that the lost Arctic voyagers were carefully selected for the service, and that each was in his condition no doubt far above the average, we will test the Esquimaux kettle-stories by some of the most trying and famous cases of hunger and exposure on record.

This, however, we must reserve for another and concluding chapter next week.

THE LOST ARCTIC VOYAGERS

Charles Dickens's rebuttal of claims made in Rae's report to the Admiralty, continued, originally published in *Household Words*, December 9, 1854.

We resume our subject of last week.

The account of the sufferings of the ship-wrecked men, in Don Juan, will rise into most minds as our topic presents itself. It is founded (so far as such a writer as Byron may choose to resort to facts, in aid of what he knows intuitively), on several real cases. Bligh's undecked-boat navigation, after the mutiny of the Bounty; and the wrecks of the Centaur, the Peggy, the Pandora, the Juno, and the Thomas; had been, among other similar narratives, attentively read by the poet.

In Bligh's case, though the endurances of all on board were extreme, there was no movement towards the "last resource." And this, though Bligh in the memorable voyage which showed his knowledge of navigation to be as good as his temper was bad (which is very high praise), could only serve out, at the best, "about an ounce of pork to each person," and was fain to weigh the allowance of bread against a pistol bullet, and in the most urgent need could only administer wine or rum by the teaspoonful. Though the necessities of the party were so great, that when a stray bird was caught, its blood was poured into the mouths of three of the people who were nearest death, and "the body, with the entrails, beak, and feet, was divided into eighteen shares." Though of a captured dolphin there was "issued about two ounces, including the offals, to each person;" and though the time came, when, in Bligh's words, "there was a visible alteration for the worse in many of the people which excited great apprehensions in me. Extreme weakness, swelled legs, hollow and ghastly countenances, with an apparent debility of understanding, seemed to me the melancholy presages of approaching dissolution."

The Centaur, man-of-war, sprung a leak at sea in very heavy weather; was perceived, after great labour, to be fast settling down by the head; and was abandoned by the captain and eleven others, in the pinnace. They were "in a leaky boat, with one of the gunwales stove, in nearly the middle of the Western Ocean; without compass, quadrant, or sail: wanting great coat or cloak; all very thinly clothed, in a gale of wind, and with a great sea running." They had "one biscuit divided into twelve morsels for breakfast, and the same for dinner; the neck of a bottle, broke off with the cork in it, served for a glass; and this filled with water was the allowance for twenty-four hours, to each man." This misery was endured, without any reference whatever to the last resource, for fifteen days: at the expiration of which time, they happily made land. Observe the captain's words, at the height. "Our sufferings were now as great as human strength could bear; but, we were convinced that good spirits were a better support than great bodily strength; for on this day Thomas Mathews, quartermaster, perished from hunger and cold. On the day before, he had complained of want of strength in his throat, as he expressed it, to swallow his morsel, and in the night grew delirious and died without a groan." What were their reflections? That they could support life on the body? "As it became next to certainty that we should all perish in the same manner in a day or two, it was somewhat comfortable to reflect that dying of hunger was not so dreadful as our imaginations had represented."

The Pandora, frigate, was sent out to Otaheite, to bring home for trial such of the mutineers of the Bounty as could be found upon the island. In Endeavour Straits, on her homeward voyage, she struck upon a reef; was got off, by great exertion; but had sustained such damage, that she soon heeled over and went down. One hundred and ten persons escaped in the boats, and entered on "a long and dangerous voyage." The daily allowance to each, was a musket-ball weight of bread, and two small wine-glasses of water. "The heat of the sun and reflexion of the sand became intolerable, and the quantity of salt water swallowed by the men created the most parching thirst; excruciating tortures were endured, and one of the men went mad and died." Perhaps this body was devoured? No. "The people at length neglected weighing their slender allowance, their mouths becoming so parched that few attempted to eat; and what was not claimed, was returned to the general stock." They were a fine crew (but not so fine as Franklin's), and in a state of high discipline. Only this one death occurred, and all the rest were saved.

The Juno, a rotten and unseaworthy ship, sailed from Rangoon for Madras, with a cargo of teak-wood. She had been out three weeks, and had already struck upon a sandbank and sprung a leak, which the crew imperfectly stopped, when she became a wreck in a tremendous storm. The second mate and others, including the captain's wife, climbed into the mizen-top, and made themselves fast to the rigging. The second mate is the narrator of their distresses, and opens them with this remarkable avowal. "We saw that we might remain on the wreck till carried off by famine, the most frightful shape in which death could appear to us. I confess it was my intention, as well as that of the rest, to prolong my existence by the only means that seemed likely to occur—eating the flesh of any whose life might terminate before my own. But this idea we did not communicate, or even hint to each other, until long afterwards; except once, that the gunner, a Roman Catholic, asked me if I thought there would be a sin in having recourse to such an expedient." Now, it might reasonably be supposed, with this beginning, that the wreck of the Juno furnishes some awful instances of the "last resource" of the Esquimaux stories. Not one. But, perhaps no unhappy creature died, in this mizen-top where the second mate was? Half a dozen, at least, died there; and the body of one Lascar getting entangled in the rigging, so that the survivors in their great weakness could not for some time release it and throw it overboard—which was their manner of disposing of the other bodies—hung there, for two or three days. It is worthy of all attention, that as the mate grew weaker, the terrible phantom which had been in his mind at first (as it might present itself to the mind of any other person, not actually in the extremity imagined), grew paler and more remote. At first, he felt sullen and irritable; on the night of the fourth day he had a refreshing sleep, dreamed of his father, a country clergyman, thought that he was administering the Sacrament to him, and drew the cup away when he stretched out his hand to take it. He chewed canvas, lead, any substance he could find—would have eaten his shoes, early in his misery, but that he wore none. And yet he says, and at an advanced stage of his story too, "After all that I suffered, I believe it fell short of the idea I had formed of what would probably be the natural consequence of such a situation as that to which we were reduced. I had read or heard that no person could live without food, beyond a few days; and when several had elapsed, I was astonished at my having existed so long, and concluded that every succeeding day must be the last. I expected, as the agonies of death approached, that we should be tearing the flesh from each other's bones." Later still, he adds: "I can give very little account of the rest of the time. The sensation of hunger was lost in that of weakness; and when I could get a supply of fresh water I was comparatively easy." When land was at last descried, he had become too indifferent to raise his head to look at it, and continued lying in a dull and drowsy state, much as Adam the interpreter lay, with Franklin at his side.

The Peggy was an American sloop, sailing home from the Azores to New York. She encountered great distress of weather, ran short of provision, and at length had no food on board, and no water, "except about two gallons which remained dirty at the bottom of a cask." The crew ate a cat they had on board, the leather from the pumps, their buttons and their shoes, the candles and the oil. Then, they went aft, and down into the captain's cabin, and said they wanted him to see lots fairly drawn who should be killed to feed the rest. The captain refusing with horror, they went forward again, contrived to make the lot fall on a negro whom they had on board, shot him, fried a part of him for supper, and pickled the rest, with the exception of the head and fingers which they threw overboard. The greediest man among them, dying raving mad on the third day after this event, they threw his body into the sea—it would seem because they feared to derive a contagion of madness from it, if they ate it. Nine days having elapsed in all since the negro's death, and they being without food again, they went below once more and repeated their proposal to the captain (who lay weak and ill in his cot, having been unable to endure the mere thought of touching the negro's remains), that he should see lots fairly drawn. As he had no security but that they would manage, if he still refused, that the lot should fall on him, he consented. It fell on a foremast-man, who was the favourite of the whole ship. He was quite willing to die, and chose the man who had shot the negro, to be his executioner. While he was yet living, the cook made a fire in the galley; but, they resolved, when all was ready for his death, that the fire should be put out again, and that the doomed foremast-man should live until an hour before noon next day; after which they went once more into the captain's cabin, and begged him to read prayers, with supplications that a sail might heave in sight before the appointed time. A sail was seen at about eight o'clock next morning, and they were taken off the wreck.

Is there any circumstance in this case to separate it from the others already described, and from the case of the lost Arctic voyagers? Let the reader judge. The ship was laden with wine and brandy. The crew were incessantly drunk from the first hour of their calamities falling upon them. They were not sober, even at the moment when they proposed the drawing of lots. They were with difficulty restrained from making themselves wildly intoxicated while the strange sail bore down to their rescue. And the mate, who should have been the exemplar and preserver of discipline, was so drunk after all, that he had no idea whatever of anything that had happened, and was rolled into the boat which saved his life.

In the case of the Thomas, the surgeon bled the man to death on whom the lot fell, and his remains were eaten ravenously. The details of this shipwreck are not within our reach; but, we confidently assume the crew to have been of an inferior class.

The useful and and accomplished Sir John Barrow, remarking that it is but too well established "that men in extreme cases *have destroyed each other* for the sake of appeasing hunger," instances the English ship the

Nautilus and the French ship the Medusa. Let us look into the circumstances of these two shipwrecks.

The Nautilus, sloop of war, bound for England with despatches from the Dardanelles, struck, one dark and stormy January night, on a coral rock in the Mediterranean, and soon broke up. A number of the crew got upon the rock, which scarcely rose above the water, and was less than four hundred yards long, and not more than two hundred broad. *On the fourth day*—they having been in the meantime hailed by some of their comrades who had got into a small whale-boat which was hanging over the ship's quarter when she struck; and also knowing that boat to have made for some fishermen not far off—these shipwrecked people ate the body of a young man who had died some hours before: notwithstanding that Sir John Barrow's words would rather imply that they killed some unfortunate person for the purpose. Now, surely after what we have just seen of the extent of human endurance under similar circumstances, we know this to be an exceptional and uncommon case. It may likewise be argued that few of the people on the rock can have eaten of this fearful food; for, the survivors were fifty in number, and were not taken off until the sixth day and the eating of no other body is mentioned, though many persons died.

We come then, to the wreck of the Medusa, of which there is a lengthened French account by two surviving members of the crew, which was very indifferently translated into English some five and thirty years ago. She sailed from France for Senegal, in company with three other vessels, and had about two hundred and forty souls on board, including a number of soldiers. She got among shoals and stranded, a fortnight after her departure from Aix Roads. After scenes of tremendous confusion and dismay, the people at length took to the boats, and to a raft made of topmasts, yards, and other stout spars, strongly lashed together. One hundred and fifty mortals were crammed together on the raft, of whom only fifteen remained to be saved at the end of thirteen days. The raft has become the ship, and may always be understood to be meant when the wreck of the Medusa is in question.

Upon this raft, every conceivable and inconceivable horror, possible under the circumstances, took place. It was shamefully deserted by the boats (though the land was within fifteen leagues at that time), and it was so deep in the water that those who clung to it, fore and aft, were always immersed in the sea to their middles, and it was only out of the water amidships. It had a pole for a mast, on which the top-gallant sail of the Medusa was hoisted. It rocked and rolled violently with every wave, so that even in the dense crowd it was impossible to stand without holding on. Within the first few hours, people were washed off by dozens, flung themselves into the sea, were stifled in the press, and getting entangled among the spars, rolled lifeless to and fro under foot. There was a cask of wine upon it which was secretly broached by the soldiers and sailors, who drank themselves so mad, that they resolved to cut the cords asunder, and send the whole living freight to perdition. They were headed by "an Asiatic, and soldier in a colonial regiment: of a colossal stature, with short curled hair, an extremely large nose, an enormous mouth, a sallow complexion, and a hideous air." Him, an officer cast into the sea; upon which, his comrades made a charge at the officer, threw *him* into the sea, and, on his being recovered by their opponents who launched a barrel to him, tried to cut out his eyes with a penknife. Hereupon, an incessant and infernal combat was fought between the two parties, with sabers, knives, bayonets, nails, and teeth, until the rebels were thinned and cowed, and they were all ferociously wild together. On *the third day*, they "fell upon the dead bodies with which the raft was covered, and cut off pieces, which some instantly devoured. Many did not touch them; almost all the officers were of this number." On the fourth "we dressed some fish (they had fire on the raft) which we devoured with extreme avidity; but, our hunger was so great, and our portion of fish so small, that we added to it some human flesh, which dressing rendered less disgusting; it was this which the officers touched for the first time. From this day we continued to use it; but we could not dress it any more, as we were entirely deprived of the means," through the accidental extinction of their fire, and their having no materials to kindle another. Before the fourth night, the raving mutineers rose again, and were cut down and thrown overboard until only thirty people remained alive upon the raft. On the seventh day, there were only twenty-seven; and twelve of these, being spent and ill, were every one cast into the sea by the remainder, who then, in an access of repentance, threw the weapons away too, all but one sabre. After that, "the soldiers and sailors" were eager to devour a butterfly which was seen fluttering on the mast; after that, some of them began to tell the stories of their lives; and thus, with grim joking, and raging thirst and reckless bathing among the sharks which had now begun to follow the raft, and general delirium and fever, they were picked up by a ship: to the number, and after the term of exposure, already mentioned.

Are there any circumstances in this frightful case, to account for its peculiar horrors? Again, the reader shall judge. No discipline worthy of the name had been observed aboard the Medusa from the minute of her weighing anchor. The captain had inexplicably delegated his authority "to a man who did not belong to the staff. He was an ex-officer of the marine, who had just left an English prison, where he had been for ten years." This man held the ship's course against the protest of the officers, who warned him what would come of it. The work of the ship had been so ill done, that even the common manœuvres necessary to the saving of a boy who fell overboard, had been bungled, and the boy had been needlessly lost. Important signals had been received from one of the ships in company, and neither answered nor reported to the captain. The Medusa had been on fire through negligence. When she struck, desertion of duty, mean evasion and fierce recrimination, wasted the precious moments. "It is probable that if one of the first officers had set the example, order would have been restored; but every one was left to himself." The most virtuous aspiration of which the soldiers were sensible, was, to fire upon their officers, and, failing that, to tear their eyes out and rend them to pieces. The historians compute that there were not in all upon the raft—before the sick were thrown into the sea—more than twenty men of decency, education, and purpose enough, even to oppose the maniacs. To crown all, they describe the soldiers as "wretches who were not worthy to wear the French uniform. They were the scum of all countries, the refuse of the prisons, where they had been collected to make up the force. When, for the sake of health, they had been made to bathe in the sea (a ceremony from which some of them had the modesty to endeavour to excuse themselves), the whole crew had had ocular demonstration that it was not upon their *breasts* these heroes wore the insignia of the exploits which had led to their serving the state in the ports of Toulon, Brest, or Rochefort." And is it with the scourged and branded sweepings of the galleys of France, in their debased condition of eight-and-thirty years ago, that we shall compare the flower of the trained adventurous spirit of the English Navy, raised by Parry, Franklin, Richardson, and Back?

Nearly three hundred years ago, a celebrated case of famine occurred in the Jacques, a French ship, homeward-bound from Brazil, with forty-five persons on board, of whom twenty-five were the ship's company. She was a crazy old vessel, fit for nothing but firewood, and had been out four months, and was still upon the weary seas far from land, when her whole stock of provisions was exhausted. The very maggots in the dust of the bread-room had been eaten up, and the parrots and monkeys brought from Brazil by the men on board had been killed and eaten, when two of the men died. Their bodies were committed to the deep. At least twenty days afterwards, when they had had perpetual cold and stormy weather, and were grown too weak to navigate the ship; when they had eaten pieces of the dried skin of the wild hog, and leather jackets and shoes, and the horn-plates of the ship-lanterns, and all the wax-candles; the gunner died. His body likewise, was committed to the deep. They then began to hunt for mice, so that it became a common thing on board, to see skeleton-men watching eagerly and silently at mouse-holes, like cats. They had no wine and no water; nothing to drink but one little glass of cider, each, per day. When they were come to this pass, two more of the sailors "died of hunger." Their bodies likewise, were committed to the deep. So long and doleful were these experiences on the barren sea, that the people conceived the extraordinary idea that another deluge had happened, and there was no land left. Yet, this ship drifted to the coast of Brittany, and no "last resource" had ever been appealed to. It is worth remarking that, *after they were saved*, the captain declared he had meant to kill somebody, privately, next day. Whosoever has been placed in circumstances of peril, with companions, will know the infatuated pleasure some imaginations take in enhancing them and all their remotest possible consequences, after they are escaped from, and will know what value to attach to this declaration.

In the reign of Queen Elizabeth, a ship's master and fifteen men escaped from a wreck in an open boat, which they weighed down very heavy, and were at sea, with no freshwater, and nothing to eat but the floating sea-weed, seven days and nights. "We will all live or die together," said the master on the third day, when one of the men proposed to draw lots—not who should become the last resource, but who should be thrown overboard to lighten the boat. On the fifth day, that man and another died. The rest were "very weak and praying for death;" but these bodies also, were committed to the deep.

In the reign of George the Third, the Wager, man-of-war, one of a squadron badly found and provided in all respects, sailing from England for South America, was wrecked on the coast of Patagonia. She was commanded by a brutal though bold captain, and manned by a turbulent crew, most of whom were exasperated to a readiness for all mutiny by having been pressed in the Downs, in the hour of their arrival at home from long and hard service. When the ship struck, they broke open the officers' chests, dressed themselves in the officers' uniforms, and got drunk in the old, Smollett manner. About a hundred and fifty of them made their way ashore, and divided into parties. Great distress was experienced from want of food, and one of the boys, "having picked up the liver of one of the drowned men whose carcase had been dashed to pieces against the rocks, could be with difficulty withheld from making a meal of it." One man, in a quarrel, on a spot which, in remembrance of their sufferings there, they called Mount Misery, stabbed another mortally, and left him dead on the ground. Though a third of the whole number were no more, chiefly through want, in eight or ten weeks; and though they had in the meantime eaten a midshipman's dog, and were now glad to feast on putrid morsels of seal that had been thrown away; certain men came back to this Mount Misery, expressly to give this body (which throughout had remained untouched), decent burial: assigning their later misfortunes "to their having neglected this necessary tribute." Afterwards, in an open-boat navigation, when rowers died at their oars of want and its attendant weakness, and there was nothing to serve out but bits of rotten seal, the starving crew went ashore to bury the bodies of their dead companions, in the sand. At such a condition did even these ill-nurtured, ill-commanded, ill-used men arrive, without appealing to the "last resource," that they were so much emaciated "as hardly to have the shape of men," while the captain's legs "resembled posts, though his body appeared to be nothing but skin and bone," and he had fallen into that feeble state of intellect that he had positively forgotten his own name.

In the same reign, an East Indiaman, bound from Surat to Mocha and Jidda in the Dead Sea, took fire when two hundred leagues distant from the nearest land, which was the coast of Malabar. The mate and ninety-five other people, white, brown, and black, found themselves in long-boat, with this voyage before them, and neither water nor provisions on board. The account of the mate who conducted the boat, day and night, is, "We were never hungry, though our thirst was extreme. On the seventh day, our throats and tongues swelled to such a degree, that we conveyed our meaning by signs. Sixteen died on that day, and almost the whole people became silly, and began to die laughing. I earnestly petitioned God that I might continue in my senses to my end, which He was pleased to grant: I being the only person on the eighth day that preserved them. Twenty more died that day. On the ninth I observed land, which overcame my senses, and I fell into a swoon with thankfulness of joy." Again no last resource, and can the reader doubt that they would all have died without it?

In the same reign, and within a few years of the same date, the Philip Aubin, bark of eighty tons, bound from Barbadoes to Surinam, broached-to at sea, and foundered. The captain, the mate, and two seamen, got clear of the wreck and into "a small boat twelve or thirteen feet long." In accomplishing this escape, they all, but particularly the captain, showed great coolness, courage, sense, and resignation. They took the captain's dog on board, and picked up thirteen onions which floated out of the ship, after she went down. They had no water, no mast, sail, or oars; nothing but the boat, what they wore, and a knife.

The boat had sprung a leak, which was stopped with a shirt. They cut pieces of wood from the boat itself, which they made into a mast; they rigged the mast with strips of the shirt; and they hoisted a pair of wide trousers for a sail. The little boat being cut down almost to the water's edge, they made a bulwark against the sea, of their own backs. The mate steered with a topmast he had pushed before him to the boat, when he swam to it. On the third day, they killed the dog, and drank his blood out of a hat. On the fourth day, the two men gave in, saying they would rather die than toil on; and one persisted in refusing to do his part in baling the boat, though the captain implored him on his knees. But, a very decided threat from the mate to steer him into the other world with the topmast by bringing it down upon his skull, induced him to turn-to again. On the fifth day, the mate exhorted the rest to cut a piece out of his thigh, and quench their thirst; but, no one stirred. He had eaten more of the dog than any of the rest, and would seem from this wild proposal to have been the worse for it, though he was quite steady again next day, and derived relief (as the captain did), from turning a nail in his mouth, and often sprinkling his head with salt-water. The captain, first and last, took only a few mouthfuls of the dog, and one of the seamen only tasted it, and the other would not touch it. The onions they all thought of small advantage to them, as engendering greater thirst. On the eighth day, the two seamen, who had soon relapsed and become delirious and quite oblivious of their situation, died, within three hours of each other. The captain and mate saw the Island of Tobago that evening, but could not make it until late in the ensuing night. The bodies were found in the boat, unmutilated by the last resource.

In the same reign still, and within three years of this disaster, the American brig, Tyrel, sailed from New York for the island of Antigua. She was a miserable tub, grossly unfit for sea, and turned bodily over in a gale of wind, five days after her departure. Seventeen people took to a boat, nineteen feet and a half long, and less than six feet and a half broad. They had half a peck of white biscuit, changed into salt dough the sea-water; and a peck of common ship-biscuit. They steered their course by the polar-star. Soon after sunset on the ninth day, the second mate and the carpenter died very peacefully. "All betook themselves to prayers, and then after some little time stripped the bodies of their two unfortunate comrades, and threw them overboard." Next night, a man aged sixty-four who had been fifty years at sea, died, asking to the last for a drop of water; next day, two more died, in perfect repose; next night, the gunner; four more in the succeeding four and twenty hours. Five others followed in one day. And all these bodies were quietly thrown overboard—though with great difficulty at last, for the survivors were new exceeding weak, and not one had strength to pull an oar. On the fourteenth or fifteenth morning, when there were only three left alive, and the body of the cabin boy, newly dead, was in the boat, the chief mate "asked his two companions whether they thought they could eat any of the boy's flesh? The signified their inclination to try; whence, the body being quite cold, he cut a piece from the inside of its thigh, a little above the knee. Part of this he gave to the captain and boatswain, and reserved a small portion to himself. But, on attempting to swallow the flesh, it was rejected by the stomachs of all, and the body was therefore thrown overboard." Yet that captain, and that boatswain both died of famine in the night, and *another whole week elapsed* before a schooner picked up the chief mate, left alone in the boat with their unmolested bodies, the dumb evidence of his story. Which bodies the crew of that schooner saw, and buried in the deep.

Only fours years ago, in the autumn of eighteen hundred and fifty, a party of British missionaries were most indiscreetly sent out by a Society, to Patagonia. They were seven in number, and all died near the coast (as nothing but a miracle could have prevented their doing), of starvation. An exploring party, under Captain Moorshead of her Majesty's ship Dido, came upon their traces, and found the remains of four of them, lying by their two boats which they had hauled up for shelter. Captain Gardiner, their superintendent, who had probably expired the last, had kept a journal until the pencil had dropped from his dying hand. They had buried three of their party, like Christian men, and the rest had faded away in quiet resignation, and without great suffering. They were kind and helpful to one another, to the last. One of the common men, just like Adam with Franklin, was "cast down at the loss of his comrades, and wandering in his mind" before he passed away.

Against this strong case in support of our general position, we will faithfully set four opposite instances we have sought out.

The first is the case of the New Horn, Dutch vessel, which was burnt at sea and blew up with a great explosion, upwards of two hundred years ago. Seventy-two people escaped in two boats. The old Dutch captain's narrative being rather obscure, and (as we believe) scarcely traceable beyond a French translation, it is not easy to understand how long they were at sea, before the people fell into the state to which the ensuing description applies. According to our calculation, however, they had not been shipwrecked many days—we take the period to have been less than week—and they had had seven or eight pounds of biscuit on board. "Our misery daily increased, and the rage of hunger urging us to extremities, the people began to regard each other with ferocious looks. Consulting among themselves, they secretly determined to devour the boys on board, and after their bodies were consumed, to throw lots who should next suffer death, that the lives of the rest might be preserved." The captain dissuading them from this with the utmost loathing and horror, they reconsidered the matter, and decided "that should we not get sight of land in three days, the boys should be sacrificed." On the last of the three days, the land was made; so, whether any of them would have executed this intention, can never be known.

The second case runs thus. In the last year of the last century, six men were induced to desert from the English artillery at St. Helena—a deserter from any honest service is not a character from which to expect much—and to go on board an American ship, the only vessel then lying in those roads. After they got on board in the dark, they saw lights moving about on shore, and, fearful that they would be missed and taken, went over the side, with the connivance of the ship's people, got into the whale boat, and made off: purposing to be taken up again by and by, when the ship was under weigh. But, they missed her, and rowed and sailed about for sixteen days, at the end of which their provisions were all consumed. After chewing bamboo, and gnawing leather, and eating a dolphin, one of them proposed, when ten days more had run out, that lots should be drawn which deserter should bleed himself to death, to support life in the rest. It was agreed to, and done. They could take very little of this food.

The third, is case of the Nottingham Galley, trading from Great Britain to America, which was wrecked on a rock called Boon Island, off the coast of Massachusetts. About two days afterwards—the narrative is not very clear in its details—the cook died on the rock. "Therefore," writes the captain, "we laid him in a convenient place for the sea to carry him away. None then proposed to eat his body, though several afterwards acknowledged that they, as well as myself, had thoughts of it." They were "tolerably well supplied with fresh-water throughout." But, when they had been upon the rock about a fortnight, and had eaten all their provisions, the carpenter died. And then the captain writes: "We suffered the body to remain with us till morning, when I desired those who were best able to remove it. I crept out myself to see whether Providence had yet sent us anything to satisfy our craving appetites. Returning before noon, and observing that the dead body still remained, I asked the men why they had not removed it: to which they answered, that all were not able. I therefore fastened a rope to it, and, giving the utmost of my assistance, we, with some difficulty, got it out of the tent. But the fatigue and consideration of our misery together, so overcame my spirits, that, being ready to faint, I crept into the tent and was no sooner there, than, as the highest aggravation of distress, the men began requesting me to give them the body of their lifeless comrade to eat, the better to support their own existence." The captain ultimately complied. They became brutalized and ferocious; but they suffered him to keep the remains on a high part of the rock: and they were not consumed when relief arrived.

The fourth and last case, is the wreck of the St. Lawrence, bound from Quebec for New York. An ensign of foot, bringing home despatches, relates how she went ashore on a desolate part of the coast of North America, and how those who were saved from the wreck suffered great hardships, both by land and sea, and were thinned in their numbers by death, and buried their dead. All this time they had some provisions, though they ran short, but at length they were reduced to live upon weeds and tallow and melted snow. The tallow being all gone, they lived on weed and snow for three days, and then the ensign came to this: "The time was now arrived when I thought it highly expedient to put the plan before mentioned (casting lots who should be killed) into execution; but on feeling the pulse of my companions, I found some of them rather averse to the proposal. The desire of life still prevailed above every other sentiment, not withstanding the wretchedness of our condition, and the impossibility of preserving it by any other method. I thought it an extraordinary instance of infatuation, that men should prefer the certainty of a lingering and miserable death, to the distant chance of escaping one more immediate and less painful. However, on consulting with the mate what was to be done, I found that although they objected to the proposal of casting lots for the victim, yet all concurred in the necessity of some one being sacrificed for the preservation of the rest. The only question was how it should be determined; when by a kind of reasoning more agreeable to the dictates of self-love than justice, it was agreed, that as the captain was now so exceedingly reduced as to be evidently the first who would sink under our present complicated misery; as he had been the person to whom we considered ourselves in some measure indebted for all our misfortunes; and further, as he had ever since our shipwreck been the most remiss in his exertions towards the general good—he was undoubtedly the person who should be the first sacrificed." The design of which the ensign writes with this remarkable coolness, was not carried into execution, by reason of their falling in with some Indians; but, some of the party who were afterwards separated from the rest, declared when they rejoined them, that they had eaten of the remains of their deceased companions. Of this case it is to be noticed that the captain is alleged to have been a mere kidnapper, sailing under false pretences, and therefore not likely to have had by any means a choice crew; that the greater part of them got drunk when the ship was in danger; and that they had not a very sensitive associate in the ensign, on his own highly disagreeable showing.

It appears to us that the influence of great privation upon the lower and least disciplined class of character, is much more bewildering and maddening at sea than on shore. The confined space, the monotonous aspect of the waves, the mournful winds, the monotonous motion, the dead uniformity of colour, the abundance of water that cannot be drunk to quench the raging thirst (which the Ancient mariner perceived to be one of his torments)—these seem to engender a diseased mind with greater quickness and of a worse sort. The conviction on the part of the sufferers that they hear voices calling for them; that they descry ships coming to their aid; that they hear the firing of guns, and see the flash; that they can plunge into the waves without injury, to fetch something or to meet somebody; is not often paralleled among suffering travellers by land. The mirage excepted—a delusion of the desert, which has its counterpart upon the sea, not included under these heads—we remember nothing of this sort experienced by Bruce, for instance, or by Mungo Park: least of all by Franklin in the memorable book we have quoted. Our comparison of the records of the two kinds of trial, leads us to believe, that even men who might be in danger of the last resource at sea, would be very likely to pine away by degrees, and never come to it, ashore.

In his published account of the ascent of Mont Blanc, which is an excellent little book, Mr. Albert Smith describes, with very humorous fidelity, that when he was urged on by the guides, in a drowsy state when he would have given the world to lie down and go to sleep for ever, he was conscious of being greatly distressed by some difficult and altogether imaginary negotiations respecting a non-existent bedstead; also, by an impression that a familiar friend in London came up with the preposterous intelligence that the King of Prussia objected to the party's advancing, because it was his ground. But, these harmless vagaries are not the present question, being commonly experienced under most circumstances where an effort to fix the attention, or exert the body, contends with a strong disposition to sleep. We have been their sport thousands of times, and have passed through a series of most inconsistent and absurd adventures, while trying hard to follow a short dull story related by some eminent conversationalist after dinner.

No statement of cannibalism, whether on the deep or the dry land, is to be admitted supposititiously, or inferentially, or on any but the most direct and positive evidence: no, not even as occurring among savage people, against whom it was in earlier times too often a pretence for cruelty and plunder. Mr. Prescott, in his brilliant history of the Conquest of Mexico, observes of a fact so astonishing as the existence of cannibalism among a people who had attained considerable advancement in the arts and graces of life, that "they did not feed on human flesh merely to gratify a brutish appetite, but in obedience to their religion—a distinction," he justly says, "worthy of notice." Besides which, it is to be remarked, that many of these feeding practices rest on the authority of narrators who distinctly saw St. James and the Virgin Mary fighting at the head of the troops of Cortes, and who possessed, therefore, to say the least, and unusual range of vision. It is curious to consider, with our general impressions on the subject—very often derived, we have no doubt, from Robinson Crusoe, if the oaks of men's beliefs could be traced back to acorns—how rarely the practice, even among savages, has been proved. The word of a savage is not to be taken for it; firstly, because he is a liar; secondly, because he is a boaster; thirdly, because he often talks figuratively; fourthly, because he is given to a superstitious notion that when he tells you he has his enemy in his stomach, you will logically give him credit for having his enemy's valour in his heart. Even the sight of cooked and dissevered human bodies among this or that tattoo'd tribe, is not proof. Such appropriate offerings to their barbarous, wide-mouthed, goggle-eyed gods, savages have been often seen and known to make. And although it may usually be held as a rule, that the fraternity of priests lay eager hands upon everything meant for the gods, it is always possible that these offerings are an exception: as at once investing the idols with an awful character, and the priests with a touch of disinterestedness, whereof their order may occasionally stand in need.

The imaginative people of the East, in the palmy days of its romance—not very much accustomed to the sea, perhaps, but certainly familiar by experience and tradition with the perils of the desert—had no notion of the "last resource" among civilized human creatures. In the whole wild circle of the Arabian Nights, it is reserved for ghoules, gigantic blacks with one eye, monsters like towers, of enormous bulk and dreadful aspect, and unclean animals lurking on the seashore, that puffed and blew their way into caves where the dead were interred. Even for Sinbad the Sailor, buried alive, the story-teller found it easier to provide some natural sustenance, in the shape of so many loaves of bread and so much water, let down into the pit with each of the other people buried alive after him (whom he killed with a bone, for he was not nice), than to invent this dismal expedient.

We are brought back to the position almost embodied in the words of Sir John Richardson towards the close of the former chapter. In weighing the probabilities and improbabilities of the "last resource," the foremost question is—not the nature of the extremity; but, the nature of the men. We submit that the memory of the lost Arctic voyagers is placed, by reason and experience, high above the taint of this so easily-allowed connection; and that the noble conduct and example of such men, and of their own great leader himself, under similar endurances, belies it, and outweighs by the weight of the whole universe the chatter of a gross handful of uncivilised people, with a domesticity of blood and blubber. Utilitarianism will protest "they are dead; why care about this?" Our reply shall be, "Because they are dead, therefore we care about this. Because they served their country well, and deserved well of her, and can ask, no more on this earth, for her justice or her loving-kindness; give them both, full measure, pressed down, running over. Because no Franklin can come back, to write the honest story of their woes and resignation, read it tenderly and truly in the book he has left us. Because they lie scattered on those wastes of snow, and are as defenceless against the remembrance of coming generations, as against the elements into which they are resolving, and the winter winds that alone can waft them home, now, impalpable air; therefore, cherish them gently, even in the breasts of children. Therefore, teach no one to shudder without reason, at the history of their end. Therefore, confide with their own firmness, in their fortitude, their lofty sense of duty, their courage, and their religion."

THE LOST ARCTIC VOYAGERS

John Rae's response to Charles Dickens, originally published in *Household Words*, December 23, 1854.

We have received the following communication from Dr. Rae. It can have no better commendation to the attention of our readers than the mention of his name:—

Observing, in the numbers of this journal dated the second and ninth of this month, a very ably-written article on the lost Arctic voyagers, in which an attempt is made to prove that Sir John Franklin's ill-fated party did not die of starvation, but were murdered by the Esquimaux; and consequently that they were not driven to the last dread alternative as a means of protracting life, permit me to make a few remarks in support of my information on this painful subject-information received by me ,with the utmost caution, and not one material point of which was published to the world without my having some good reason to support it.

First, as regards my interpreter. To compare either Augustus or Ouligback (who accompanied Sir John Franklin and Sir John Richardson in their overland journeys) with William Ouligback, my interpreter, would be very unfair to the latter. Neither of the first two could make themselves understood in the English language, and did not very perfectly comprehend the dialect of the natives of the coast westward of the Coppermine River. William Ouligback speaks English fluently; and, perhaps, more correctly than one half of the lower classes in England or Scotland.

As I could not, from my ignorance of the Esquimaux tongue, test William Ouligback's qualifications, I resorted to the only means of doing so I possessed. There is an old servant of the company at Churchill, an honest, trustworthy man, who has acquired a very fair knowledge of both the Esquimaux character and the Esquimaux language. This man informed me that young Ouligback could be perfectly relied on; that he would tell the Esquimaux exactly what was said, and give the Esquimaux reply with equal correctness; that when he had any personal object to gain, he would not scruple to tell a falsehood to attain it, but in such a case the untruth was easily discovered by a little cross-questioning. This description I found perfectly true.

Again: the natives of Repulse Bay speak precisely the same language as those of Churchill, where young Ouligback was brought up.

The objection offered that my information was received second-hand, I consider much in favour of its correctness. Had it been obtained from the natives who had seen the dead bodies of our countrymen, I should have doubted all they told me, however plausible their tale might have appeared; because had they, as they usually do, deposited any property under, stones in the neighbourhood, they would have had a very excellent cause for attempting to mislead me.

That ninety-nine out of a hundred interpreters are under a strong temptation to exaggerate, may be true. If so, my interpreter is the exception, as he did not like to talk more than he could possibly help. No doubt had I offered him a premium for using his tongue freely he might have done so; but not even the shadow of a hope of a reward was held out.

It is said that part of the information regarding cannibalism was conveyed to me by gestures. This is another palpable mistake, which is likely to mislead. I stated in one of my letters to the Times that the natives had preceded me to Repulse Bay; and, by signs, had made my men left in charge of the property there (none of whom spoke a word of Esquimaux) comprehend what I had already learnt through the interpreter.

I do not infer that the officer who lay upon his double-barrelled gun defended his life to the last against ravenous seamen; but that he was a brave, cool man, in the full possession of his mental faculties to the last; that he lay down in this position as a precaution, and, alas! was never able to rise again; and that he was among the last, if not the very last, of the survivors.

The question is asked, was there any fuel in that desolate place for cooking the contents of the kettles? I have already mentioned in a letter to the Times how fuel might have been obtained. I shall repeat my opinion with additions:—When the Esquimaux were talking with me on the subject of the discovery of the men, boats, tents, &c., several of them remarked that it was curious no sledges were found at the place. I replied that the boat was likely fitted with sledge-runners that screwed on to it. The natives answered, that sledges were noticed with the party of whites when alive, and that their tracks on the ice and snow were seen near the place where the bodies were found. My answer then was, That they must have burnt them for fuel; and I have no doubt but that the kegs or cases containing the ball and shot must have shared the same fate.

Had there been no bears thereabouts to mutilate those bodies—no wolves, no foxes? is asked; but it is a well-known fact that, from instinct, neither bears, wolves, nor foxes, nor that more ravenous of all, the glutton or wolverine, unless on the verge of starvation, will touch a dead human body; and the carnivorous quadrupeds near the Arctic sea are seldom driven to that extremity.

Quoting again from the article on the lost Arctic voyagers. "Lastly, no man can with any show of reason undertake to affirm that the sad remnant of Franklin's gallant band were not set upon and slain by the Esquimaux themselves?"

This is a question which like many others is much more easily asked than answered; yet I will give my reasons for not thinking, even for a moment, that some thirty or forty of the bravest class of one of the bravest nations in the world, even when reduced to the most wretched condition, and having firearms and ammunition in their hands, could be overcome by a party of savages equal in number to themselves. I say equal in number, because the Esquimaux to the eastward of the Coppermine, seldom, if ever, collect together in greater force than thirty men, owing to the difficulty of obtaining the means of subsistence. When Sir John Ross wintered three years in Prince Regent's Inlet, the very tribe of Esquimaux who saw Sir John Franklin's party were constantly or almost constantly in the neighbourhood. In the several springs he passed there, parties of his men were travelling in various directions; yet no violence was offered to them, although there was an immense advantage to be gained by the savages in obtaining possession of the vessels and their contents.

In eighteen hundred and forty-six-seven I and a party of twelve persons wintered at Repulse Bay. In the spring my men were divided and scattered in all directions; yet no violence was offered, although we were surrounded by native families, among whom there were at least thirty men. By murdering us they would have put themselves in possession of boats and a quantity of cutlery of great value to them. In the same spring, when perfectly alone and unarmed, except with a common clasp knife, which could have been of no use, I met on the ice four Esquimaux armed with spear and bow and arrow. I went up to them, made them shake hands; and, after exchanging a few words and signs, left them. In this case no violence was used; although I had a box of astronomical instruments on my back, which might have excited their cupidity. Last spring, I, with seven men, was almost in constant communication with a party four times our number. The savages made no attempt to harm us. Yet wood, saws, daggers, and knives were extremely scarce with them, and by getting possession of our boat, its masts and oars, and the remainder of our property, they would have been independent for years.

What appears to me the most conclusive reason for believing the Esquimaux report, is this: the natives of Repulse Bay, although they visit and communicate for mutual advantage with those further west, both dislike and fear their neighbours, and not without cause; as they have behaved treacherously to them on one or two occasions. So far do they carry this dislike, that they endeavoured, by every means in their power, to stimulate me to shoot several visitors to Repulse Bay, from Pelly Bay, and from near Sir John Ross's wintering station in Prince Regent's Inlet.

Now, is it likely that, had they possessed such a powerful argument to excite—as they expected to do—my anger and revenge as the murder of my countrymen, would they not have made use of it by acquainting me with the whole circumstances, if they had any such to report?

Again, what possible motive could the Esquimaux have for inventing such an awful tale as that which appeared in my report to the secretary of the Admiralty. Alas! these poor people know too well what starvation is, in its utmost extremes, to be mistaken on such a point. Although these uneducated savages—who seem to be looked upon by those who know them not, as little better than brutes—resort to the "last resource" only when driven to it by the most dire necessity. They will starve for days before they will even sacrifice their dogs to satisfy the cravings of their appetites.

One or two facts are worth a hundred theories on any subject. On meeting some old acquaintances among the natives at Repulse Bay, last spring, I naturally enquired about others that I had seen there in eighteen hundred and forty-six and forty-seven. The reply was, that many of them had died of starvation since I left, and some from a disease which, by description, resembled influenza. Among the party that died of starvation was one man whom I well knew—Shi-makeck—and for whom I enquired by name. I learnt that this man, rather than endure the terrible spectacle of his children pining away in his presence, went out and strangled himself. Another, equally well known to me, being unable, I suppose, to support the pangs of hunger, stripped off his clothes, and exposed himself to cold, until he was frozen to death. In several instances, on this occasion, cannibalism had been resorted to, and two women were pointed out to me as having had recourse to this "last resource." It may be, I have only the words of "babbling and false savages who are, without exception, in heart, covetous, treacherous, and cruel," in support of what I say.

Let us enquire slightly into that want of truthfulness so frequently and indiscriminately charged against savages in general, and the Esquimaux in particular:—When that most distinguished of Arctic navigators—Sir Edward Parry—wintered at Winter Island, not Winter Harbour, and at Igloolik, in the Straits of the Fury and Hecla, he met many of the very tribe of Esquimaux that I saw at Repulse Bay. From these Sir Edward received information and tracings of the coast west of Melville Peninsula, surrounding a bay named by the natives—Akkoolee.

This Esquimaux tracing or delineation of coast was entered in the charts in dotted lines, until my survey of eighteen hundred and forty-seven showed that, in all material points, the accounts given by the natives were perfectly correct. When Sir John Ross wintered three years in Prince Regent's Inlet, the natives drew charts of the coast line to the southward of his position, and informed him that, in that direction, there was no water communication leading to the western sea.

Sir John Ross's statements, founded on those of the natives were not believed at the Admiralty, nor my own, in eighteen hundred and forty-seven, although *I saw the land all the way*, and in which I was supported by Esquimaux information. The authorities at the Admiralty would still have Boothia an Island. Last spring I proved beyond the possibility of a doubt, the correctness of my former report, and consequently the truthfulness of the Esquimaux; for, where parties of high standing at home would insist on having nothing but salt water, I travelled over a neck of land or isthmus only sixty miles broad.

On conversing with the natives about the different parties of whites, and the ships and boats they had seen, they described so perfectly the personal appearance of Sir John Ross and Sir James Ross—although the men spoken with had not seen these gentlemen—that anyone acquainted with these officers could have recognised them. The natives on one point set me right, when they thought I had made a mistake. I told them that the two chiefs (Sir J. and Sir J.C. Ross) and their men had all got home safe to their own country. They immediately remarked, "that this was not true, for some of the men had died at the place where the vessel was left." I, of course, alluded only to that portion of the party who had got away from Regent's Inlet in safety. It must be remembered that this circumstance occurred upwards of twenty years ago, and consequently is an instance of correctness of memory and truthfulness that would be considered surprising among people in an advanced state of civilisation.

The peculiarities of the Great Fish River, and of the coast near its mouth, has been so minutely described by Sir George Back, and so beautifully illustrated by his admirable drawings, that they can easily be understood by any one. The Esquimaux details on this subject agreed perfectly with those of Sir George Back: the river was described as full of falls and rapids, and that many Esquimaux dwelt on or near its banks.

They described the land about a long day's journey (which, with dogs and sledges, is from thirty-five to forty miles) to the north-west of the mouth of the river, as low and flat, without hills of any kind, agreeing in every particular with the descriptions of Sir George Back and Simpson.

They told me that the top of the cairn erected by Dease and Simpson at the Castor and Pollux River had fallen down. This I found to be true; and afterwards, on asking them in which direction it had fallen, they said towards the east. True again. I showed two men, who said they had been along the coast which I had traced, my rough draft of a chart. They immediately comprehended the whole; examined and recognised the several points, islands, &c., laid down upon it; gave me their Esquimaux names, showed me where they had had "caches;" which I actually saw.

Another Esquimaux, on learning that we had opened a "cache," in which we found a number of wings and heads of geese which had lain long there, and were perfectly de-nuded of flesh, said that the "cache" belonged to him. Thinking that he was stating a falsehood so as to obtain some reward for having interfered with his property, I produced my chart, and told him to show me the island, among a number of similar ones all small, on which his "cache" was; he, without a moment's hesitation, pointed to the right island.

Having dwelt thus much on the trustworthiness of the Esquimaux, I shall next touch on their disposition and aptitude to falsehood; but this I must defer for the present.

[Charles Dickens interrupts] We will merely append, as a commentary on the opinion of our esteemed friend, Dr. Rae, relative to the probabilities of the Esquimaux besetting a forlorn and weak party, the speciality of whose condition that people are quite shrewd enough to have perceived; an extract from Sir John Barrow's account of Franklin's and Richardson's second journey:—

Thus far all went on well; but an accident happened while the crowd was pressing round the boats, which was productive of unforeseen and very annoying consequences:

"A kaiyack being overset by one of the Lion's oars, its owner was plunged into the water with his head in the mud, and apparently in danger of being drowned. We instantly extricated him from his unpleasant situation, and took him into the boat until the water could be thrown out of his kaiyack; and Augustus, seeing him shivering with cold, wrapped him up in his own great coat. At first he was exceedingly angry, but soon became reconciled to his situation; and, looking about, discovered that we had many bales, and other articles in the boat, which had been concealed from the people in the kaiyacks, by the coverings being carefully spread over all. He soon began to ask for everything he saw, and expressed much displeasure on our refusing to comply with his demands; he also, we afterwards learned, excited the cupidity of others by his account of the inexhaustible riches in the Lion, and several of the younger men endeavoured to get into both our boats, but we resisted all their attempts."

They continued, however, to press, and made many efforts to get into the boats, while the water had ebbed so far that it was not knee-deep at the boats, and the younger men, waiting in crowds around them, tried to steal everything they could reach. The Reliance being afloat, was dragged by the crowd towards the shore, when Franklin directed the crew of the Lion (which was aground and immoveable) to endeavour to follow her, but the boat remained fast until the Esquimaux lent their aid and dragged her after the Reliance. One of the Lion's men perceived that the man who was upset had a pistol under his shirt, which it was discovered had been stolen from Lieutenant Back, and the thief, seeing it to be noticed, leaped out of the boat and joined his countrymen, carrying with him the great coat which Augustus had lent him.

"Two of the most powerful men, jumping on board at the same time, seized me by the wrists and forced me to sit between them; and as I shook them loose two or three times, a third Esquimaux took his station in front to catch my arm whenever I attempted to lift my gun, or the broad dagger which hung by my side. The whole way to the shore they kept repeating the word 'teyma,' beating gently on my left breast with their hands, and pressing mine against their breasts. As we neared the beach, two oomiaks, full of women, arrived, and the 'teymas' and vociferation were redoubled. The Reliance was first brought to the shore, and the Lion close to her a few seconds afterwards. The three men who held me now leaped ashore, and those who remained in their canoes, taking them out of the water, carried them to a little distance. A numerous party then drawing their knives, and stripping themselves to the waist, ran to the Reliance, and having first hauled her as far up as they could, began a regular pillage, handing the articles to the women, who, ranged in a row behind, quickly conveyed them out of sight."

In short, after a furious contest, when knives were brandished in a most threatening manner, several of the men's clothes cut through, and the buttons of others torn from their coats, Lieutenant Back ordered his people to seize and level their muskets, but not to fire till the word was given. This had the desired effect, the whole crowd taking to their heels and hiding themselves behind the drift-timber on the beach. Captain Franklin still thought it best to temporise so long as the boats were lying aground, for armed as the Esquimaux were with long knives, bows, arrows, and spears, fire-arms could not have been used with advantage against so numerous a host; Franklin, indeed, states his conviction, "considering the state of excitement to which they had worked themselves, that the first blood which his party might unfortunately have shed, *would instantly have been revenged by the sacrifice of all their lives.*"

As soon as the boats were afloat and making to a secure anchorage, seven or eight of the natives walked along the beach, entered into conversation with Augustus, and invited him to a conference on shore. "I was unwilling to let him go," says Franklin, "but the brave little fellow entreated so earnestly that I would suffer him to land and reprove the Esquimaux for their conduct, that I at length consented." On his return, being desired to tell what he had said to them, "he had told them," he said—

"Your conduct has been very bad, and unlike that of all other Esquimaux. Some of you even stole from me, your countryman; but that I do not mind,—I only regret that you should have treated in this violent manner the white people, who came solely to do you kindness. My tribe were in the same unhappy state in which you now are, before the white people came to Churchill, but at present they are supplied with everything they need, and you see that I am well clothed; I get all that I want, and am very comfortable. Yon cannot expect, after the transactions of this day, that these people will ever bring goods to your country again, unless you show your contrition by restoring the stolen goods. The white people love the Esquimaux, and wish to show them the same kindness that they bestow upon the Indians. *Do not deceive yourselves, and suppose they are afraid of you*; I tell you they are not; and that it is entirely owing to their humanity that many of you were not killed to-day; for they have all guns, with which they can destroy you either when near or at a distance. I also have a gun, and can assure you, that if a white man had fallen, I would have been the first to have revenged his death."

The language of course is that of Franklin, who however gives it as the purport of Augustus's speech, and adds, "his veracity is beyond all question with the party." "We could perceive," says Franklin, "by the shouts of applause, with which they filled the pauses in his language, that they assented to his *arguments*;" [that is, to his representation of the superior power of those white men]; "and he told us they had expressed great sorrow for having given so much cause of offence." He said, moreover, that they pleaded ignorance, having never before seen white men; that they had seen so many fine things entirely new to them, that they could not resist the temptation of stealing; they promised never to do the like again; and gave a proof of their sincerity by restoring the articles that had been stolen. And thus in an amicable manner was the affray concluded.

DR. RAE'S REPORT

John Rae's response to Charles Dickens, continued, originally published in *Household Words*, December 30, 1854.

Dr. Rae's communication to us on the subject of his Report, which was begun last week, resumes and concludes as follows:

When the Esquimaux have an object to gain, they will not hesitate to tell a falsehood, but they cannot lie with a good grace; "they cannot lie like truth," as civilised men do. Their fabrications are so silly and ridiculous, and it is so easy to make them contradict themselves by a slight cross-questioning, that the falsehood is easily discovered. I could give a number of instances of this, but shall confine myself to two.

When Sir John Richardson descended the McKenzie in 1848, a great number of Esquimaux came off in their canoes; they told us that on an island to which they pointed, a number of white people had been living for some time; that they had been living there all winter, and that we ought to land to see them. Their story was altogether so incredible, that we could not have a moment's doubt or difficulty in tracing its object. They wished to get us on shore in order to have a better opportunity of pillaging our boats, as they did those of Sir John Franklin; for it must be remembered that the Esquimaux at the McKenzie and to the westward are different from any of those to the eastward. The former, notwithstanding the frequent efforts of the Hudson's Bay Company to effect a peace, are at constant war with the Louchoux Indians, and consequently with the "white men," as they think the latter, by supplying guns and ammunition to the Louchoux, are their allies.

Another instance excited much interest in England when it was first made known here. It was reported to Captain McClure by an Esquimaux, that one of a party of white men had been killed by one of his tribe near Point Warren. That the white men built a house there, but nobody knew how they came, as they had no boat; and that they went inland. When asked "when this took place?" the reply was, that "it might be last year or when I was a child."

How anyone could place any faith in such a report as this, I am at a loss to discover. Any man at all acquainted with the native character, would in a moment set down this tale at its proper value; at least Sir John Richardson and I did—and the first is high authority. Indeed, throughout the whole of Captain or Commander McClure's communication with the natives in the neighbourhood of the McKenzie, he appears to have been admirably imposed upon by them. Let us again get at a fact or two.

He is told by a chief that the Esquimaux go so far to the westward to trade, instead of to the McKenzie, "because, at the latter place, the white man had given the Indians very bad water, which killed many and made others foolish (drunk), and that they would not have any such water. From this it evidently appears that the Company lose annually many valuable skins, which find their way to the Colvill instead of to the McKenzie."

Let us quietly examine the above statements. It is well known that since the McKenzie has been discovered, ardent spirits have not been admitted within the district, for the natives. At present, and for many years back spirits or wines have not been allowed to enter the McKenzie or its neighbouring district of Athabasca, as allowances for either officers or men in the Hudson's Bay Company's service, so that the natives might not have it to say that we took for ourselves what we would not give to them. We do not know, nor do I think that there are, any Russian trading posts on the Colvill. The true reason that these

Esquimaux do not trade with the Hudson's Bay Company is, that the former are constantly at war with the Louchoux. Frequent attempts have been made to effect a reconciliation between these tribes, but hitherto without success.

Captain McClure tells us that the Esquimaux informed him that "they had no communication with any person belonging to the Great River" (McKenzie); yet, strange to say, he intrusts the very despatches in which this is mentioned, to natives of the same tribe, and indulges the hope that his "letter may reach the Hudson's Bay Company this year," (one thousand eight hundred and fifty). In another case, Captain McClure mentions that he gave a gun and ammunition to an Esquimaux chief, to deliver a despatch into the hands of the Hudson's Bay Company. In any case, prepayment is acknowledged to be a bad plan, but worst of all in that of a savage with whom you are unacquainted, and on whom you have no hold. Had the pay depended upon the performance of the service, the despatch might have had some chance of reaching its destination.

I have had some opportunities of studying Esquimaux character; and, from what I have seen, I consider them superior to all the tribes of red men in America. In their domestic relationship they show a bright example to the most civilised people. They are dutiful sons and daughters, kind brothers and sisters, and most affectionate parents. So well is the first of these qualities understood among them, that a large family is considered wealth by a father and mother—for, the latter well know that they will be carefully tended by their offspring, well clothed and fed, whilst a scrap of skin or a morsel of food is to be obtained, as long as a spark of life remains; and, after death, that their bodies will be properly placed either on or under the ground, according to the usage of the tribe.

I do not stand alone in the high opinion I have formed of the Esquimaux character. At the Hudson's Bay Company's establishments of Fort George on the east, and Churchill on the west, coast of Hudson's Bay, where the Esquimaux visit, they are looked upon in an equally favourable light. The Moravian missionaries on the Labrador coast find the Esquimaux honest and trustworthy, and employ them constantly and almost exclusively as domestic servants. The report of the residents in the Danish settlements on the west shores of Greenland, is no less favourable; and although I have no special authority for saying so, I believe that Captain Perring's opinions are similar. During the two winters I passed at Repulse Bay, I had men with me who had been, at some time of their lives, in all parts of the Hudson's Bay Company's territories. These men assured me that they had never seen Indians so decorous, obliging, unobtrusive, orderly, and friendly, as the Esquimaux.

Oh! some one may remark, perhaps they have some private reason for this.

Now, my men had not any "private reason" for saying so. I firmly believe, and can almost positively assert, that no case of improper intercourse took place between them and the natives of Repulse Bay during the two seasons I remained there—which is more, I suspect, than most of the commanders of parties to the Arctic Sea can truthfully affirm. A number of instances (principally shipwrecks), are brought forward to show that cannibalism has not been usually resorted to in cases of extreme want; that it is the exception, not the rule. Yet not one of those properly represent the probable position of Sir John Franklin's party. In all the cases above alluded to, the parties suffering were deprived of water as well as of food. We all know that when anyone suffers from two painful sensations, but painful in different degrees, the more severe of the two prevents the lesser from being felt.

Thirst causes a far more painful sensation than hunger, and consequently, whilst the first remains unappeased, the pangs of the other are very slightly, if at all, felt. In the case of Franklin's party, their thirst could be easily assuaged, and consequently the pangs of hunger would be felt the more intensely. Even Franklin's former disastrous journey (from the narrative of which large extracts have been made) is not a parallel case. In it the suffering party had generally something or other every few days to allay the cravings of hunger. They had pieces of old leather, tripe de roche, and an infusion of the tea-plant. Unfortunately, near the mouth of Back's Fish River, there are none of the above named plants,—nothing but a barren waste with scarcely a blade of grass upon it. Much stress is laid on the moral character and the admirable discipline of the crews of Sir John Franklin's ships. What their state of discipline may have been I cannot say, but their conduct at the very last British port they entered was not such as to make those who knew it, consider them very deserving of the high eulogium passed upon them in Household Words. Nor can we say that the men, in extreme cases of privation, would maintain that state of subordination so requisite in all cases, but more especially during danger and difficulty.

We have, I am sorry to say, but too many recent instances of disagreement and differences among the officers employed on the Arctic service. It is well known in naval circles that, in one vessel which has not yet arrived from the north, there will be two or three courts martial as soon as she reaches home. To place much dependence on the obedience and good conduct of the comparatively uneducated seamen, if exposed to the utmost extremes of distress, when their superiors, without having any such excuse, have forgotten themselves on a point of such vital importance, would be very unreasonable. Besides, seamen generally consider themselves, when they have lost their ship and set foot on shore, as being freed from that strict discipline to which they would readily submit themselves when on board.

As these observations have already attained a much greater length than I at first anticipated, I shall refrain from mentioning, as I intended, one or two instances of persons fully as well educated as the generality of picked seamen usually are, and brought up as Christians, having, in cases of extreme want, had recourse to the "last resource," as a means of maintaining life.

I am aware of the difficulties I have to encounter in replying to the article on the "Lost Arctic Voyagers." That the author of that article is a writer of very great ability and practice, and that he makes the best use of both to prove his opinions, is very evident. Besides, he takes the popular view of the question, which is a great point in his favour. To oppose this, I have nothing but a small amount of practical knowledge of the question at issue, with a few facts to support my views and opinions; but, I can only throw them together in a very imperfect and unconnected form, as I have little experience in writing, and, like many men who have led a wandering and stirring life, have a great dislike to it. It is seldom that a man can do well what is disagreeable to him.

That my opinions remain exactly the same as they were when my report to the Admiralty was written, may be inferred from all I have now stated.

That twenty or twenty-five Esquimaux could, for two months together, continue to repeat the same story without variation in any material point, and adhere firmly to it, in spite of all sorts of cross-questioning, is to me the clearest proof that the information they gave me was founded on fact.

That the "white men" were not murdered by the natives, but that they died of starvation, is, to my mind, equally beyond a doubt.

In conclusion, let me remark, that I fully appreciate the kind, courteous, and flattering manner in which my name is mentioned by the writer on the subject of the lost Arctic Voyagers.

SIR JOHN FRANKLIN
AND HIS CREWS

John Rae's full report to the Hudson's Bay Company, dated September 1, 1854, printed in *Household Words*, May 1, 1855.

In order that our readers, at a future time, when the Esquimaux stories shall have been further tested, may be in possession of them as originally brought home, we have procured from Dr. Rae a faithful copy of his Report for publication. We do not feel justified in omitting or condensing any part of it; believing, as we do, that it is a very unsatisfactory document on which to found such strong conclusions as it takes for granted. The preoccupation of the public mind has dismissed this subject easily for the present; but, we assume its great interest, and the serious doubts we hold of its having been convincingly set at rest, to be absolutely certain to revive.

York Factory, Hudson's Bay, 1st Sept., 1854.
I have the honour to report, for the information of the Governor, Deputy Governor, and Committee, that I arrived here yesterday with my party, all in good health; but, from causes which will be explained hereafter, without having effected the object of the expedition. At the same time information has been obtained, and articles purchased from the natives, which prove beyond a doubt that a portion, if not all, of the survivors of the long lost and unfortunate party under Sir John Franklin had met with a fate as melancholy and dreadful as it is possible to imagine.

By a letter dated Chesterfield Inlet, ninth of August, eighteen hundred and fifty-three, you are in possession of my proceedings up to that time. Late on the evening of that day we parted company with our small consort, she steering down to the south ward, whilst we took the opposite direction to Repulse Bay.

Light and variable winds sadly retarded our advance northward; but by anchoring during the flood, and sailing or rowing with the tide, we gained some ground daily. On the eleventh we met with upwards of three hundred walrus, lying on a rock a few miles off shore. They were not at all shy, and several were mortally wounded, but one only (an immensely large fellow) was shot dead by myself. The greater part of the fat was cut off and taken on board, which supplied us abundantly with oil for our lamps all winter.

On the forenoon of the fourteenth, having a fair wind, we rounded Cape Horn, and ran up Repulse Bay; but as the weather was very foggy, completely hiding every object at the distance of a quarter-of-a-mile, we made the land about seven miles east of my old winter quarters; next day, midst heavy rain, we ran down to North Pole River, moored the boat, and pitched the tents.

The weather being still dark and gloomy, the surrounding country presented a most dreary aspect. Thick masses of ice clung to the shore, whilst immense drifts of snow filled each ravine, and lined every steep bank that had a southerly exposure. No Esquimaux were to be seen, nor any recent traces of them. Appearances could not be less promising for wintering safely; yet I determined to remain until the first of September; by which date some opinion could be formed as to the practicability of procuring sufficient food and fuel for our support during the winter: all the provisions on hand at that time being equal to only three months' consumption.

The weather fortunately improved, and not a moment was lost. Nets were set; hunters were sent out to procure venison; and the majority of the party was constantly employed collecting fuel. By the end of August a supply of the latter essential article (Andromeda Tetragona) for fourteen weeks was laid up, thirteen deer and one musk-bull had been shot, and one hundred and thirty-six salmon caught. Some of the favourite haunts of the Esquimaux had been visited, but no indications were seen to lead us to suppose that they had been lately in the neighbourhood.

The absence of the natives caused me some anxiety; not that I expected any aid from them, but because I could attribute their having abandoned so favourable a locality to no other cause than a scarcity of food, arising from the deer having taken another route in their migrations to and from the north.

On the first of September I explained our position to the men; the quantity of provisions we had, and the prospects, which were far from flattering, of getting more. They all most readily volunteered to remain, and our preparations for a nine months' winter were continued with unabated energy. The weather, generally speaking, was favourable, and our exertions were so successful, that by the end of the month we had a quantity of provisions and fuel collected adequate to our wants up to the period of the spring migrations of the deer.

One hundred and nine deer, one musk-ox (including those killed in August) fifty-three brace of ptarmigan, and one seal, had been shot; and the nets produced fifty-four salmon. Of the larger animals above enumerated, forty-nine deer and the musk-ox were shot by myself; twenty-one deer by Mistegan, the deer-hunter; fourteen by another of the men; nine by William Ouligback; and sixteen by the remaining four men.

The cold weather set in very early, and with great severity. On the twentieth, all the smaller, and some of the larger lakes, were covered with ice four to six inches thick. This was far from advantageous for deer shooting, as these animals were enabled to cross the country in all directions, instead of following their accustomed passes.

October was very stormy and cold. About the fifteenth, the migrations of the deer terminated, and twenty-five more were added to our stock. Forty-two salmon, and twenty trout, were caught with nets and hooks set in lakes under the ice. On the twenty-eighth, the snow was packed hard enough for building; and we were glad to exchange the cold and dismal tents (in which the temperature had latterly been thirty-six or thirty-seven degrees below the freezing point) for the more comfortable shelter of snow-houses, which were built on the south south-east side of Beacon Hill, by which they were well protected from the prevailing north-west gales. The houses were nearly half a mile south of my winter quarters of eighteen hundred and forty-six and eighteen hundred and forty-seven.

The weather in November was comparatively fine, but cold, the highest, lowest, and mean temperature being, respectively, thirty-eight degrees, eighteen degrees, and three degrees below zero. Some deer were occasionally seen, but only four were shot; some wolves, several foxes, and one wolverine were killed; and from the nets fifty-nine salmon and twenty-two trout were obtained.

Our most productive fishery was in a lake about three miles distant, bearing east (magnetic) from Beacon Hill, or the mouth of the North Pole River.

The whole of December, a very few days excepted, was one continued gale with snow and drift. When practicable, the men were occupied scraping under snow for fuel, by which means our stock of that very essential article was kept up. The mean temperature of the month was twenty-three degrees below zero. The produce of our nets and guns was extremely small, amounting to one partridge, one wolf, and twenty-seven fish.

On the first of January, eighteen hundred and fifty-four, the temperature rose to the very unusual height: of eighteen degrees above zero, the wind at the time being south-east, with snow. Our nets, after being set in different lakes without success, were finally taken up on the twelfth, only five small fish having been caught. The thermometer was tested by freezing mercury, and found to be in error, the temperature indicated by it being four degrees five minutes too high.

The cold during February was steady and severe, but there were fewer storms than usual. Deer were more numerous, and generally were travelling northward. One or two were wounded, but none killed. On two occasions (the first and twenty-seventh), that beautiful but rare appearance of the clouds near the sun, with three fringes of pink and green, following the outline of the cloud, was seen, and I may add that the same splendid phenomenon was frequently observed during the spring, and was generally followed by a day or two of fine weather.

During the latter part of the month, preparations were being made for our spring journeys. A carpenter's workshop was built of snow, and our sledges were taken to pieces, reduced to as light a weight as possible, and then reunited more securely than before. The mean temperature of February, corrected for error of thermometer, was thirty-nine degrees below zero. The highest and lowest being twenty degrees and fifty-three degrees.

On the first of March a female deer in fine condition was shot, and on the ninth and tenth two more were killed. Three men were absent some days during this month, in search of Esquimaux, from whom we wished to obtain dogs. They went as far as the head of Ross Bay, but found no traces of these people.

On the fourteenth I started with three men hauling sledges with provisions, to be placed in "cache" for the long spring journey. Owing to the stormy state of the weather we got no farther than Cape Lady Pelly, on the most northerly point of which our stores were placed, under a heap of large stones, secure from any animal except man or the bear. We returned on the twenty-fourth, the distance walked together being a hundred and seventy miles.

On the thirty-first of March, leaving three men in charge of the boat and stores, I set out with the other four, including the interpreter, with the view of tracing the west coast of Boothia, from the Castor and Pollux River to Bellot Strait. The weight of our provisions, &c., with those deposited on the way, amounted to eight hundred and sixty-five pounds, an ample supply for sixty-five days.

The route followed for part of the journey being exactly the same as that of spring, eighteen hundred and forty-seven, it is unnecessary to describe it. During the two first days, although we did not travel more than fifteen miles per day, the men found the work extremely hard, and as I perceived that one of them (a fine, active young fellow, but a light weight) would be unable to keep pace with the others, he was sent back, and replaced by Mistegan, a very able man, and an experienced sledge-hauler. More than a day was lost in making this exchange, but there was still abundance of time to complete our work, if not opposed by more than common obstacles.

On the sixth of April we arrived at our provision cache, and found it all safe. Having placed the additional stores on the sledges, which made those of the men weigh more than a hundred and sixty pounds each, and my own about a hundred and ten pounds, we travelled seven miles further, then built a snow house on the ice two miles from shore. We had passed among much rough ice, but hitherto the drift banks of snow, by lying in the same direction in which we were travelling, made the walking tolerably good. As we advanced to the northward, however, these crossed our track (showing that the prevailing winter gales had been from the westward), and together with stormy weather, impeded us so much that we did not reach Colville Bay until the tenth. The position of our snow house was in latitude sixty-eight degrees thirteen minutes five seconds north, longitude by chronometer eighty-eight degrees fourteen minutes fifty-one seconds west, the variation of the compass being eighty-six degrees twenty minutes west. From this place it was my intention to strike across land as straight as possible for the Castor and Pollux River.

The eleventh was so stormy that we could not move, and the next day, after placing en cache two days provisions, we had walked only six miles in a westerly direction, when a gale of wind compelled us to get under shelter. The weather improved in the evening, and having the benefit of the full moon, we started again at a few minutes to eight P.M. Our course at first was the same as it had been in the morning, but the snow soon became so soft and so deep that I turned more to the northward in search of firmer footing. The walking was excessively fatiguing, and would have been so even to persons travelling unencumbered, as we sank at every step, nearly ankle deep in snow. Eight and a half miles were accomplished in six and a half hours, at the end of which as we required some rest, a small snow house was built, and we had some tea and frozen pemmican.

After resting three hours we resumed our march, and by making long detours, found the snow occasionally hard enough to support our weight. At thirty minutes to noon on the thirteenth, our day's journey terminated in latitude sixty-eight degrees twenty-three minutes thirty seconds north, longitude eighty-nine degrees three minutes fifty-three seconds west, variation of compass eighty-three degrees thirty minutes west. At a mile and a half from our bivouac, we had crossed the arm of a lake of considerable extent, but the country around was so flat, and so completely covered with snow, that its limits could not be easily defined, and our snow hut was on the borders of another lake apparently somewhat smaller.

A snowstorm of great violence raged during the whole of the fourteenth, which did not prevent us from making an attempt to get forward. After persevering two and a half hours, and gaining a mile and a half distance, we were again forced to take shelter.

The fifteenth was very beautiful, with a temperature of only eight degrees below zero. The heavy fall of snow had made the walking and sledge-hauling worse than before. It was impossible to keep a straight course, and we had to turn much out of our way, so as to select the hardest drift banks. After advancing several miles, we fortunately reached a large lake containing a number of islands, on one of which I noticed an old Esquimaux tent site. The fresh footmarks of a partridge (Tetrao rupestris) were also seen, being the only signs of living thing (a few tracks of foxes excepted) that we had observed since commencing the traverse of this dreary waste of snow-clad country. To the lake above mentioned, and to those seen previously, the name of Barrow was given, as a mark of respect to John Barrow, Esquire, of the Admiralty; whose zeal in promoting, and liberality in supporting, many of the expeditions to the Arctic Sea are too well known to require any comment, further than that he presented a very valuable Halkett's boat for the service of my party, which unfortunately by some irregularity in the railway baggage trains between London and Liverpool did not reach the latter place in time for the steamer, although sent from London some days before. Our snow hut was built on the edge of a small lake in latitude sixty-eight degrees thirty-one minutes thirty-eight seconds north, longitude eighty-nine degrees eleven minutes fifty-five seconds west, variation of compass eighty-three degrees thirty minutes west.

The difficulties of walking were somewhat diminished on the sixteenth by a fresh breeze of wind, which drifted the snow off the higher ground, and we were enabled to make a fair day's journey. Early on the seventeenth we reached the shore of Pelly Bay, but had barely got a view of its rugged ice covering before a dense fog came on. We had to steer by compass for a large rocky island, some miles to the westward; and we stopped on an islet near its east shore until the fog cleared away. This luckily happened some time before noon, and afforded an opportunity of obtaining observations, the results of which were latitude sixty-eight degrees forty-four minutes fifty-three seconds north, longitude by chronometer eighty-nine degrees thirty-four minutes forty-seven seconds west, and variation eighty-four degrees twenty minutes west.

Even on the ice we found the snow soft and deep, a most unusual circumstance. The many detentions I had met with caused me now, instead of making for the Castor and Pollux River, to attempt a direct course towards the magnetic pole, should the land west of the bay be smooth enough for travelling over. The large island west of us was so rugged and steep that there was no crossing it with sledges; we therefore travelled along its shores to the northward, and stopped for the night within a few miles of the northern extremity. The track of an Esquimaux sledge drawn by dogs was observed to-day, but it was of old date.

The morning of the eighteenth was very foggy; but after rounding the north point of the island it became clear, and we travelled due west, or very nearly so, until within three miles of the west shore of the bay, which presented an appearance so rocky and mountainous, that it was evident we could not traverse it without loss of time. As the country towards the head of the bay looked more level, I turned to the southward, and, after a circuitous walk of more than sixteen miles, we built our snow house on the ice, five miles from shore. Many old traces of Esquimanx were seen on the ice to-day.

On the nineteenth we continued travelling southward, and our day's journey (about equal to that of yesterday) terminated near the head of the bay.

Twentieth of April. The fresh footmarks of Esquimaux, with a sledge, having been seen yesterday on the ice within a short distance of our resting-place, the interpreter and one man were sent to look for them, the other two being employed in hunting and collecting fuel, whilst I obtained excellent observations, the results of which were latitude sixty-eight degrees twenty-eight minutes twenty-nine seconds north, longitude by chronometer ninety degrees eighteen minutes thirty-two seconds west, variation of compass ninety-eight degrees thirty minutes west. The latter is apparently erroneous, probably caused by much local attraction.

After an absence of eleven hours the men sent in search of Esquimaux returned in company with seventeen natives (five of whom were women), and several of them had been at Repulse Bay when I was there in eighteen hundred and forty-seven. Most of the others had never before seen "whites," and were extremely forward and troublesome. They would give us no information on which any reliance could be placed, and none of them would consent to accompany us for a day or two, although I promised to reward them liberally.

Apparently, there was a great objection to our travelling across the country in a westerly direction. Finding that it was their object to puzzle the interpreter and mislead us, I declined purchasing more than a small piece of seal from them, and sent them away—not, however, without some difficulty, as they lingered about with the hope of stealing something; and, notwithstanding our vigilance, succeeded in abstracting from one of the sledges a few pounds of biscuit and grease.

The morning of the twenty-first was extremely fine; and at three A.M. we started across land towards a very conspicuous hill, bearing west of us. On a rocky eminence, some miles inland, we made a cache of the seal's flesh we had purchased. Whilst doing this, our interpreter made an attempt to join his countrymen. Fortunately, his absence was observed before he had gone far; and he was overtaken after a sharp race of four or five miles. He was in a great fright when we came up to him, and was crying like a child, but expressed his readiness to return, and pleaded sickness as an excuse for his conduct. I believe he was really unwell—probably from having eaten too much boiled seal's flesh, with which he had been regaled at the snow huts of the natives.

Having taken some of the lading off Ouligback's sledge, we had barely resumed our journey when we were met by a very intelligent Esquimaux, driving a dog-sledge laden with musk-ox beef. This man at once consented to accompany us two days' journey, and in a few minutes had deposited his load on the snow, and was ready to join us. Having explained my object to him, he said that the road by which he had come was the best for us; and, having lightened the men's sledges, we travelled with more facility.

We were now joined by another of the natives, who had been absent seal-hunting yesterday; but being anxious to see us had visited our snow-house early this morning, and then followed our track. This man was very communicative, and on putting to him the usual questions as to his having seen white men before, or any ships or boats, he replied in the negative; but said that a party of kabloonans had died of starvation a long distance to the west of where we then were, and beyond a large river. He stated that he did not know the exact place—that he had never been there, and that he could not accompany us so far.

The substance of the information then and subsequently obtained from various sources was to the following effect.

In the spring, four winters past (eighteen hundred and fifty), whilst some Esquimaux families were killing seals near the northern shore of a large island, named in Arrowsmith's charts King William Land, about forty white men were seen travelling in company southward over the ice, and dragging a boat and sledges with them. They were passing along the west shore of the above-named island. None of the party could speak the Esquimaux language so well as to be understood; but by signs the natives were led to believe that the ship or ships had been crushed by ice, and that they were then going to where they expected to find deer to shoot. From the appearance of the men—all of whom, with the exception of an officer, were hauling on the drag-ropes of the sledge, and were looking thin—they were then supposed to be getting short of provisions; and they purchased a small seal, or piece of seal, from the natives. The officer was described as being a tall, stout, middle-aged man. When their day's journey terminated, they pitched tents to rest in.

At a later date, the same season, but previous to the disruption of the ice, the corpses of some thirty persons and some graves were discovered on the continent, and five dead bodies on an island near it, about a long day's journey to the north-west of the mouth of a large stream, which can be no other than Back's Great Fish River (named by the Esquimaux Oot-koo-hi-ca-lik), as its description, and that of the low shore in the neighbourhood of Point Ogle and Montreal Island, agree exactly with that of Sir George Back. Some of the bodies were in a tent or tents; others were under the boat, which had been turned over to form a shelter; and some lay scattered about in different directions. Of those seen on the island, it was supposed that one was that of an officer (chief), as he had a telescope strapped over his shoulders, and his double-barrelled gun lay underneath him.

From the mutilated state of many of the bodies, and the contents of the kettles, it is evident that our wretched countrymen had been driven to the last dread alternative as a means of sustaining life.

A few of the unfortunate men must have survived until the arrival of the wild fowl (say until the end of May), as shots were heard, and fishbones and feathers of geese were noticed near the scene of the sad event.

There appears to have been an abundant store of ammunition, as the gunpowder was emptied by the natives in a heap on the ground out of the kegs or cases containing it; and a quantity of shot and ball was found below high-water mark, having probably been left on the ice close to the beach before the spring thaw commenced. There must have been a number of telescopes, guns (several of them double-barrelled), watches, compasses, &c.; all of which seem to have been broken up, as I saw pieces of these different articles with the natives,—and I purchased as many as possible, together with some silver spoons and forks, an order of merit in the form of a star, and a small silver plate engraved "Sir John Franklin, K.C.H."

Enclosed is a list of the principal articles bought, with a note of the initials, and a rough pen-and-ink sketch of the crests on the forks and spoons. The articles themselves I shall have the honour of handing over to you on my arrival in London.

None of the Esquimaux with whom I had communication saw the white men, either when living or after death, nor had they ever been at the place where the corpses were found, but had their information from natives who had been there, and who had seen the party when travelling over the ice. From what I could learn, there is no reason to suspect that any violence had been offered to the sufferers by the natives.

As the dogs in the sledge were fatigued before they joined us, our day's journey was a short one. Our snow-house was built in latitude sixty-eight degrees twenty-nine seconds north, and longitude ninety degrees forty-two minutes forty-two seconds west, on the bed of a river having high mud banks, and which falls into the west side of Pelly Bay, about latitude sixty-eight degrees forty-seven minutes north, and longitude ninety degrees twenty-five minutes west.

On the twenty-second, we travelled along the north bank of the river (which I named after Captain Beecher, of the Admiralty), in a westerly direction, for seven or eight miles, until abreast of the lofty and peculiarly shaped hill already alluded to, and which I named Ellice Mountain, when we turned more to the northward.

We soon arrived at a long narrow lake, on which we encamped a few miles from its east end,—our day's march being little more than thirteen miles. Our Esquimaux auxiliaries were now anxious to return, being in dread, or professing to be so, that the wolves or wolverines would find their "cache" of meat, and destroy it. Having paid them liberally for their aid and information, and having bade them a most friendly farewell, they set out for home as we were preparing to go to bed.

Next morning provisions for six days were secured under a heap of ponderous stones, and we resumed our march along the lake.

Thick weather, snow-storms, and heavy walking, sadly retarded our advance. The Esquimaux had recommended me, after reaching the end of the chain of lakes (which ran in north-westerly direction for nearly twenty miles, and then turned sharply to the southward) to follow the windings of a brook that flowed from them. This I attempted to do, until finding that we should be led thereby far to the south, we struck across land to the west among a series of hills and valleys.

Tracks of deer now became numerous, and a few traces of musk cattle were observed.

At two A.M., on the twenty-sixth, we fell upon a river with banks of mud and gravel twenty to forty feet high, and about a quarter of a mile in width. After a most laborious walk of more than eighteen miles, we found an old snow-hut, which after a few repairs was made habitable, and we were snugly housed at forty minutes past six A.M. Our position was in latitude sixty-eight degrees twenty-five minutes twenty-seven seconds north, longitude ninety-two degrees fifty-three minutes fourteen seconds west.

One of our men who, from carelessness some weeks before, had severely frozen two of his toes, was now scarcely able to walk; and as, by Esquimaux report, we could not be very far from the sea, I prepared to start in the evening with two men and four days' provisions for the Castor and Pollux River, leaving the lame man and another to follow, at their leisure a few miles on our track, to some rocks that lay on our route where they were more likely to find both fuel and game, than on the bare flat ground where we then were.

The morning of the twenty-sixth was very fine as we commenced tracing the course of the river seaward; sometimes following its course, at other times travelling on its left or right bank to cut off points.

At four A.M., on the twenty-seventh, we reached the mouth of the river, which, by subsequent observation, I found to be situated in latitude sixty-eight degrees thirty-two minutes north, and longitude ninety-three degrees twenty minutes west. It was rather difficult to discover when we had reached the sea, until a mass of rough ice settled the question beyond a doubt. After leaving the river we walked rapidly due west for six miles, then built our usual snug habitation on the ice, three miles from shore, and had some partridges (Tetrao mutus) for supper, at the unseasonable hour of eight A.M. We had seen great numbers of these birds during the night.

Our latitude was sixty-eight degrees thirty-two minutes one second north, and about forty minutes east of Simpson's position of the mouth of the Castor and Pollux River.

The weather was overcast with snow when we resumed our journey, at thirty minutes past eight P.M., on the twenty-seventh; we directed our course directly for the shore, which we reached after a sharp walk of one and a half hours, in doing which we crossed a long stony island of some miles in extent. As by this time it was snowing heavily, I made my men travel on the ice, the walking being better there, whilst I followed the winding of the shore, closely examining every object along the beach.

After passing several heaps of stones, which had evidently formed Esquimaux caches, I came to a collection larger than any I had yet seen, and clearly not intended for the protection of property of any kind. The stones, generally speaking, were small, and had been built in the form of a pillar, but the top had fallen down, as the Esquimaux had previously given me to understand was the case.

Calling my men to land, I sent one to trace what looked like the bed of a small river immediately west of us, whilst I and the other man cleared away the pile of stones in search of a document. Although no document was found, there could be no doubt in my own mind, and in that of my companion, that its construction was not that of the natives. My belief that we had arrived at the Castor and Pollux River was confirmed when the person who had been sent to trace the apparent stream-bed returned with the information that it was a river.

My latitude of the Castor and Pollux is sixty-eight degrees twenty-eight minutes thirty-seven seconds, west; agreeing within a quarter of a mile with that of Simpson; but our longitudes differ considerably, his being ninety-four degrees fourteen minutes west, whilst mine was ninety-three degrees forty-two minutes west. My longitude is nearly intermediate between that of Simpson and Sir George Back, supposing the latter to have carried on his survey eastward from Montreal Island. A number of rocky elevations to the north of the river were mistaken by Simpson for islands, and named by him the Committee.

Having spent upwards of an hour in fruitless search for a memorandum of some kind we began to retrace our steps; and after a most fatiguing march of fifteen hours, during which we walked at least thirty miles, we arrived at the snow-hut of the men left behind. They had shot nothing, and had not collected sufficient andromeda for cooking, but had been compelled to use some grease. The frost-bitten man could scarcely move.

Early on the morning of the twenty-ninth, during a heavy fall of snow, we set out for the mouth of the river, which was named in honour of Sir Frederick Murchison, the late President of the Royal Geographical Society; and after losing our way occasionally in attempting to make short cuts, we arrived at Cache Island, so named from an Esquimaux cache that was on it, within two miles of the sea, at eight A.M., and stopped there, as it blew a gale with drift.

As soon as we got shelter, and had supped, preparations were made for starting in the evening for Bellot Strait. An ample stock of provisions and fuel for twenty-two days were placed on two of our best sledges, and I hauled on my own small sledge my instruments, books, bedding, &c., as usual.

On the evening of the twenty-ninth, the weather was so stormy, that although we were prepared to start at eight o'clock, we could not get away until past two on the following morning, when after travelling little more than five miles, a heavy fall of snow and strong wind caused us again to take shelter.

Our advance was so much impeded by thick weather and soft snow, that we did not arrive within a few miles of Cape Porter of Sir John Ross, until the sixth of May. In doing this we had traversed a bay, the head of which was afterwards found to extend as far north as latitude sixty-eight degrees four minutes north. Point Sir H. Dryden, its western boundary, is in latitude sixty-eight degrees forty-four minutes north, longitude ninety-four degrees west. To this bay, the name of Shepherd was given, in honour of the Deputy Governor of the Honourable Hudson's Bay Company, and an island near its head, was called Bence Jones, after the distinguished medical man and analytical chemist of that name to whose kindness I and my party were much indebted, for having proposed the use and prepared some extract of tea, for the expedition.

This article we found extremely portable, and as the tea could be made without boiling water, we often enjoyed a cup of that refreshing beverage, when otherwise from want of fuel, we must have been satisfied with cold water.

From Point Dryden, the coast which is low and stony, runs in a succession of small points and bays about ten miles nearly due west, then turns sharply up to the north in latitude sixty-eight degrees forty-five minutes north, longitude ninety-four degrees twenty-seven minutes fifty seconds west, which was ascertained by observations obtained on an island near the shore. The point was called Cape Colville, after the Governor of the Company, and the island, Stanley. To the west, at the distance of seven or eight miles, land was seen, which received the appellation of Matheson Island, as a mark of respect to one of the Directors of the Company.

Our snow-hut on the sixth of May, situate on Pointe de la Guiche was by good observations found to be in latitude sixty-eight degrees fifty-seven minutes fifty-two seconds north, longitude ninety-four degrees twenty-two minutes fifty-eight seconds west. One of my men, Mistegan, an Indian of great intelligence and activity, was sent six miles farther along the coast northwards; by ascending some rough ice at its extreme point, he could see about five miles farther, the land was still trending northward, whilst to the north-west, at a considerable distance, perhaps twelve or fourteen miles, there was an appearance of land, the channel between which and the point where he stood, being full of rough ice. This land, if it was such, is probably part of Matty Island, or King William Land, which latter is also clearly an island.

I am happy to say that on this present, as on a former, occasion, where my survey met that of Sir James C. Ross, a very singular agreement exists, considering the circumstances under which our surveys have been taken.

The foggy and snowy weather, which continued upwards of four days, had occasioned the loss of so much time, that, although I could easily have completed a part (perhaps the half) of the survey of the coast, between the Magnetic Pole and Bellot Strait, or Brentford Bay, I could not do the whole without great risk to my party, and I therefore decided upon returning.

Having taken possession of our discoveries in the usual form, and built a cairn, we commenced our return on the night of the sixth. Having fine, clear weather, we made long marches, and at Shepherd Bay, having got rid of the sledge, which I had hitherto hauled, I detached myself from the party, and examined the bay within a mile or two of the shore, whilst my men took a straighter route.

Thick weather again came on as we entered the bay (named in honour of Sir Robert H. Inglis) into which the Murchison River falls, and we had much trouble in finding the mouth of the river. Here the services of my Cree hunter were of much value, as custom had caused him to notice indications and marks, which would have escaped the observation of a person less acute and experienced.

On the eleventh of May, at three A.M., we reached the place where our two men had been left. Both were as well as I could hope for, the one whose great toe had been frozen, and which was about to slough off at the first joint, thereby rendering the foot very tender and painful when walking in deep snow, had too much spirit to allow himself to be hauled. One deer, and eighteen partridges had been shot; but, notwithstanding, I found a greater reduction in our stock of provisions than I had anticipated, and I felt confirmed in the course I had taken.

The day became very fine, and observations were taken, which gave the position of Cache Island, where our snow-hut was—latitude sixty-eight degrees thirty-two minutes two seconds north, longitude ninety-three degrees thirteen minutes eighteen seconds west.

Having completed my observations, and filled in rough tracings of the coast line, which I generally did from day to day, we started for home at eight thirty P.M. The weather being now fine, and the snow harder than when outward bound, we advanced more rapidly and in a straighter direction, until we came to the lakes, about midway in the Isthmus, after which, as far as Pelly Bay, our outward and homeward route were exactly alike. We reached Pelly Bay at one A.M., on the seventeenth, and built a snow-house about two and a half miles south, and the same distance west, of my observations of the twentieth of April.

Observing traces of Esquimaux, two men were sent, after supper, to look for them. After eight hours absence they returned with ten or twelve native men, women, and children. From these people I bought a silver spoon and fork. The initials F.R.M.C., not engraved, but scratched with a sharp instrument, on the spoon, puzzled me much, as I knew not at the time the Christian names of the officers of Sir John Franklin's expedition; and thought that the letters above-named might possibly be the initials of Captain McClure, the small c between M C being omitted.

Two of the Esquimaux (one of them I had seen in eighteen hundred and forty-seven) offered for a consideration to accompany us a day or two's march with a sledge and dogs. We were detained some time by the slow preparations of our new allies; but we soon made up for lost time, and, after a journey of sixteen geographical or about eighteen and a half statute miles, we arrived at the east side of the bay, in latitude by reduction to the meridian sixty-eight degrees twenty-three minutes ten seconds north, longitude eighty-nine degrees fifty-eight minutes thirty-nine seconds west.

It may be remembered that in the spring of eighteen forty-seven I did not trace the shore of Pelly Bay, but saw it from the summit of one of the lofty islands in the bay. Desirous of being always within, rather than of exceeding the limits of truth, I that year placed the head of the bay about ten miles north of what it ought to have been,—a mistake which will be easily accounted for by those who know the difficulties of estimating distances in a snow-clad country, where the height of the land is unknown.

The width of the isthmus separating Pelly and Shepherd's Bays is fully sixty geographical miles.

In the evening before parting with our Esquimaux assistants, we bought a dog from them, and after a most friendly farewell, resumed our journey eastward, and found, on a long lake, some old snow-houses, in which we took up our lodgings. Here a set of good observations placed us in latitude sixty-eight degrees twelve minutes eighteen seconds north, longitude eighty-nine degrees twenty-four minutes fifty-one seconds west; variation eighteen-one degrees west.

On the morning of the twenty-first, we arrived at Committee Bay. From thence our route to Repulse Bay was almost the same as before; and I shall not, therefore, advert to it further than to mention that we arrived at our winter home at five, A.M., on the twenty-sixth of May,—having, from the better walking, travelled in twenty days the distance (less forty or fifty miles) which had taken us thirty-six days to accomplish on our outward journey.

I found the three men who had been left in charge of the property quite well, living in abundance, and on the most friendly terms with a number of Esquimaux families, who had pitched their tents near them.

The natives had behaved in the most exemplary manner; and many of them who were short of food, in compliance with my orders to that effect, had been supplied with venison from our stores.

It was from this time until August that I had opportunities of questioning the Esquimaux regarding the information which I had already obtained, of the party of whites who had perished of starvation, and of eliciting the particulars connected with that sad event, the substance of which I have already stated.

In the early part of July, the salmon came from the sea to the mouths of the rivers and brooks which were at that date open; and we caught numbers of them. So that occasionally we could afford to supply our native friends with fifty or one hundred in a night. As is the usual custom at the Hudson's Bay Company's inland trading posts, all provisions were given gratis; and they were much more gratefully received by the Esquimaux than by the more southerly and more favoured red man.

We had still on hand half of our three months' stock of pemmican, and a sufficiency of ammunition to provide for the wants of another winter. We were all in excellent health, and could get as many dogs as we required: so that (D.V.) there was little doubt that a second attempt to complete the survey would be successful; but I now thought that I had a higher duty to attend to, that duty being to communicate, with as little loss of time as possible, the melancholy tidings which I had heard, and thereby save the risk of more valuable lives being jeopardised in a fruitless search, in a direction where there was not the slightest prospect of obtaining any information. I trust this will be deemed a sufficiently good reason for my return.

The summer was extremely cold and backward; we could not leave Repulse Bay until the fourth of August, and on the sixth had much difficulty in rounding Cape Hope. From thence, as far as Cape Fullerton, the strait between Southampton Island and the main shore was fully packed with ice, which gave us great trouble. South of Cape Fullerton we got into open water. On the evening of the nineteenth instant, calms and head winds much retarded us, so that we did not enter Churchill River until the morning of the twenty-eighth of August. There we were detained all day by a storm of wind. My good interpreter, William Ouligback, was landed, and before bidding him farewell, I presented him with a very handsomely mounted hunting knife, intrusted to me by Captain Sir George Back for his former travelling companion, Ouligback; but as the old man was dead, I took the liberty of giving it to his son, as an inducement to future good conduct should his services be again required.

A three days' run brought us to York Factory, at which place we landed all well on the forenoon of the 31st of August. I am happy to say that the conduct of my men, under circumstances often very trying, was generally speaking extremely good and praiseworthy; and although their wages were higher than those of any party who have hitherto been employed on boat expeditions, I thought it advisable, after consulting with Chief Factor William Mactavish, to give each a small gratuity, varying the amount according to merit.

In conclusion, I have to express my regret that I was unable, on this occasion, to bring to a successful termination an expedition which I had myself planned and projected; but in extenuation of my failure, I may mention that I was met by an accumulation of obstacles, beyond the usual ones of storms and rough ice, which my former experience in Arctic travelling had not led me to anticipate.

ABOUT JOHN RAE

JOHN RAE was a Scottish doctor and explorer born on September 30, 1813, in Orkney. As a child, he enjoyed sailing, climbing, hunting, and fishing—skills that would serve him well in his future exploits. In 1833, shortly after graduating from medical school, Rae was appointed surgeon of the *Prince of Wales*, a Hudson's Bay Company ship bound for Moose Factory, Ontario, where he remained for the next ten years.

Rae had great respect for the peoples native to northern Canada and adopted many of their survival skills. He learned to hunt caribou, store meat, construct shelter, and walk using snowshoes. He was particularly known for this last skill, once walking 1,200 miles in snowshoes through winter forest in order to learn how to survey.

Rae embarked on his first expedition in 1846. Over the next decade, he explored much of northern Canada's coastline and, in fact, discovered the final link in the Northwest Passage. However, his achievements have gone largely unrecognized due to the discovery he made in 1854 regarding the fate of the Franklin expedition and the subsequent criticism he received upon his return to England.

John Rae retired from the Hudson's Bay Company in 1856 but retained his love of exploration. In 1860, he was hired to explore Iceland and Greenland in an effort to establish a telegraph line to America. And in 1884, at the age of seventy-one, he was hired by the HBC to survey another telegraph route, in the west of Canada, from Red River to Victoria.

John Rae died in London on July 22, 1893. He was the only major explorer of his time not to receive a knighthood.

ABOUT KEN McGOOGAN

The award-winning author of ten books, KEN McGOOGAN is best known for *Fatal Passage: The Untold Story of John Rae, the Arctic Explorer Who Discovered the Fate of Franklin*. That work won the Drainie-Taylor Biography Prize, the CAA History Award, and an American Christopher Award for "a work of artistic excellence that affirms the highest values of the human spirit." With his related book, *Lady Franklin's Revenge*, Ken added the UBC Medal for Canadian Biography and the Pierre Berton Award for History. He writes a column for *Canada's History* magazine, recently published *How the Scots Invented Canada*, and teaches a creative non-fiction course through the University of Toronto and the New York Times Knowledge Network. Ken also makes a cameo appearance in the acclaimed docudrama *Passage*, which is based on *Fatal Passage*. Please visit www.kenmcgoogan.blogspot.com.